Books by Sue Hubbell

FAR FLUNG HUBBELL

A BOOK OF BEES

A COUNTRY YEAR

ON THIS HILLTOP

BROADSIDES FROM THE
OTHER ORDERS

WAITING FOR APHRODITE

SUE HUBBELL

Waiting for
Aphrodite

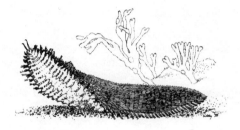

JOURNEYS INTO THE
TIME BEFORE BONES

With illustrations by Liddy Hubbell

HOUGHTON MIFFLIN COMPANY
BOSTON • NEW YORK
1999

For information about permission to reproduce selections
from this book, write to Permissions, Houghton Mifflin Company,
215 Park Avenue South, New York, New York 10003.

Library of Congress Cataloging-in-Publication Data
Hubbell, Sue.
Waiting for Aphrodite / Sue Hubbell ; with illustrations
by Liddy Hubbell.
p.   cm.
Includes bibliographical references (p.   ) and index.
ISBN 0-395-83703-0
1. Invertebrates. I. Title.
QL362.H835 1999
592 — dc21   98-49811 CIP

Book design by Robert Overholtzer

Printed in the United States of America

QUM 10 9 8 7 6 5 4 3 2 1

Portions of Chapter 9 were originally published in the *New York Times,*
June 15, 1997. Copyright © 1997 by the New York Times Co. Reprinted by permission. The poems by Ralph A. Lewin on pages 34
and 80, from *The Biology of Algae and Diverse Other Verses* (Boxwood
Press, 1987), are reprinted by permission. The poem by Brian Hubbell
on page 18 is reprinted by permission of the author.

FOR LOUIE,
who wanted to be in the picture

THE NEW NOTEBOOK

Full of superstition
I begin a new notebook,
white leaves — sea foam.
I close my eyes and wait
for the first day of the world,
for Aphrodite with wet lips,
red curls of flame,
an open shell,
shy and sure,
to rise from the salt foam,
out of the primordial algae.
I wait under closed eyelids.
One can hear the grey rustle of sea gulls,
under the low sky
and the monotonous thunder of waves
only of waves
which come and go.

> — Maria Banus
>   *Translated from Romanian by*
>   *Laura Schiff and Dana Beldiman*

Acknowledgments

THIS BOOK was shaped by scientists and others who generously agreed to share their knowledge and expertise with me and kindly agreed to review the manuscript when it was written. I am deeply in their debt, but any mistakes in the book are mine, not theirs.

Other people also contributed in important ways. No matter what troubles, fiscal and organizational, the Library of Congress suffers through, the wonderful librarians there become enthusiastically involved in research. In particular, I should like to single out one of the science librarians, Allison V. Level, for literally going the extra mile. Shirley Newman, at the M. L. King Public Library, in Washington, also became involved to a far greater degree than any writer should expect. I am grateful to Michael Sieverts, who provided materials when I was temporarily library-less.

Laurie Snyder and John Wood, who have seeing eyes, spotted Aristotle's lantern lying by its lone, and for that piece of magic I am grateful. John Campbell, the affable classicist, has often helped me with writing projects, and he figures prominently in this one. Lee Hudson and her family, proprietors of Frenchman's Bay Fishery, found sea cucumbers and sea mice for me; my thanks to them. And I am grateful to Jessie Just for asking the question that put the book into focus for me and to Asher Treat for his sharp intelligence and understanding of a good phrase.

Liz Darhansoff gave me a career that I enjoyed, and Joe Fox, who during his lifetime never wished to be mentioned, patted it along. My gratitude goes to both of them, as it does to Nancy Newhouse, an editor who was always willing to take a chance. My thanks also to Harry Foster, an editor with a portable generator, who showed up in my life at just the right moment.

Arne, whose many virtues include a master's touch on the photocopying machine, I thank for the support he gave and continues to give. And to LBN, who wanted to stay out of the picture, I am grateful for everything else.

WAITING
FOR
APHRODITE

Prologue

FOR TWENTY-FIVE YEARS I lived on a ninety-acre farm in the Ozarks of southern Missouri. For a number of those years I lived alone after my first marriage ended. But in the mid-1980s I married an old friend, Arne, a man who worked and lived in Washington, D.C., and we bought a house there that is even older than we are. I began commuting regularly, every couple of months, between the two places. But not in the summertime. Arne and I share a number of crotchets that make young people look amused: we don't watch television, and we don't have a fax machine, computer, cellular telephone, or microwave oven. And, most important to my point here, our house lacks air conditioning. Summers in Washington are, if anything, more humid and hotter than they are in southern Missouri, so we began renting summer places in New England. In antique Victorian fashion I would take up residence there, and he (who had an air-conditioned office to escape to) would join me when his work permitted.

I have lots of connections to New England, some created during the five years that I lived and worked in Rhode Island. And I have been visiting Maine for the last fifty years, had grown to appreciate the year-round beauty and interest of the land along its coast. But it wasn't until just a few years ago that I began to think seriously about making it my home.

Missouri had been home for a long time. My first husband and I had driven the seven miles of bad, rocky, rutted road to the farm for the first time with a real estate agent, and at the end of the driveway we had found a dilapidated cabin and a rundown farm that somehow, in its blowsy, generous beauty, I'd known would be home. During the twenty-five years I lived there I created a beekeeping and honey business, turned the place into something that was more nature park than farm, and restored the worn-out land. But the work was hard, and sixty is not thirty-five. At sixty, Missouri farmers begin to make plans to sell the old homestead and move to town. I was not interested in moving to town because the years had brought development to it. Plans were in the works to bring a four-lane divided highway through, the final connection of a southern superhighway between the east and west coasts. In a few years my Missouri town would be a service area.

I had returned to one particular summer cottage on the coast of Maine for several years before I realized that one reason I kept coming back was that the nearby town was a Yankee version of what my Missouri town had been when I had first moved there, before it had traffic lights, before Wal-Mart and McDonalds stripped its downtown. I decided that this town in Maine was the place I wanted to grow old in.

During the course of my visits to Maine, I must have driven by the house on the hill, with its weathered FOR SALE sign, hundreds of times. The place had been vacant and available for so many years that it had become invisible and, in fact, the little house could not be seen from the road. But one hot afternoon, at the end of two desultory months spent with real estate agents looking at all the wrong properties along the Maine coast, I *did* see the sign as I was driving back to the rented cottage. I was tired and discouraged, but the dogs, Tazzie and Louie, needed to get out of the car anyway, so, with a sigh, I turned up the crumbling blacktop driveway and saw the disappointing house: ugly, boxy, a 1950s ranch trying to make more of itself than it should have. I let the dogs out to run around in the untended yard, where iris and lilies were blooming among the wildflowers. The dogs checked the place out, tails wagging, and I sat down under the shade of a maple tree that, unaccountably, fit my back perfectly. I looked east toward the next day's sunrise and saw the sea. A breeze was blowing, and I could smell salt, seaweed, and sun-bleached shore. I knew, once again, that I'd found home. Houses can be fixed.

I was surprised that this strange little house was going to be my home, so I got up from under the maple tree, whistled up the dogs, and decided to look around. Together we scrambled up over the ledge in back of the house, crossed the moss- and lichen-covered boulders, walked through the groves of pine, spruce, maple, birch, and fir that grew where there was soil enough to sustain them. The blueberries were not yet ripe but soon would be. From the bald, rocky crown of the hill I could see the entire ocean bay, the islands lying in it, and the sea shining beyond. Climbing back down, I found a rock-encircled spring shaded by trees. In another age it would have had its very own naiad. Louie thought it was lovely and threw herself in for a drink. But she was startled to find it so deep that she had to swim, and scrabbled out onto the rocks with difficulty. Back at the house I tried to peer through the windows, but the blinds were tightly drawn.

A week later a real estate agent let me inside. It was bad. The rooms were low-ceilinged and cramped. The windows were small, with light showing around the edges of the jambs, which meant they'd be drafty in the winter. A sump pump ran nearly full time, removing standing water from the cellar. Lime green shag carpeting covered the floors throughout. Bad. Bad. Bad. But not *too* bad. The carpets probably hid perfectly decent hardwood flooring. I could rip off the roof, build up, top the house off with the glass tower I'd always wanted. Knock out some ceilings. Knock out some walls. Ungirdle the place. Allow it to sprawl a little. Perhaps some porches? Replace the windows. The freestanding shop could be turned into a guest house, for the people who always come to visit you in Maine. Not bad at all. I knew that this time I didn't have twenty-five years to build everything myself, the way I had done in Missouri. But I knew an architect. I knew a contractor. I had a wrecking bar.

I phoned Arne in Washington and told him I had found a place that I thought would work out. He flew up for a weekend. Arne has made do a great deal in his life, but he has never ripped off roofs or knocked down walls. He was also educated at Oxford, where he acquired formidable rhetorical skills. He looked around the house, then turned to me and said, "Well, it is just AWFUL, but if we buy it I can't see that there's that much wrong with it."

Buy it we did. We hired an architect and a contractor, and during the summer I camped out in a house in process. The plumber whom the contractor hired is a wise man. He knew the place; he'd done work for the previous owners. He agreed with me that the house needed improvement and said he'd enjoy helping. "It's such a pretty spot," he said. "What if you'd found a really nice house in an ugly place? Now that is what would be bad."

All that summer, backhoes ditched for drainage and new foundations. Trucks moved earth from here to there. Concrete was poured. I applied my wrecking bar. I smashed all the pink and gray ceramic tile in the bathroom with a sledgehammer and shoveled it out. Arne flew up weekends and grimaced at the piles of dirt and fill. He winced as the plumber slowly and noisily drilled a hole for the guest-house plumbing, through what Arne said was "a perfectly good wall." But I showed him the spring, and before long he was cutting a pretty walking trail up to it. He explored on his own and found the USGS marker at the peak of our tiny mountain. He found a cove where the seawater was warm enough to swim in. He found the best place to buy fresh-picked crabmeat. Arne made do, making, in his way, a home too.

Adding Maine to my life was easy. It happened gradually, over a number of years of visiting, summer renting, buying a place, beginning work on it, making friends. It all seemed natural, fitting, proper.

But subtracting Missouri from my life was hard. I knew with my head that the time had come to let my place go. The hard physical work of managing a large bee farm in an isolated area miles from neighbors at the end of a bad road seemed more difficult as each year passed. Arne and I, over the last few years, had been taken up with concern and care for aging relatives and friends, and from them I had learned that there are good and bad ways to grow old. The details differ in each case, but the bad ways have a number of similarities: a rigid, gripping conservatism, leading to a fearfulness of new things, new people, new places, and a reluctance to leave a way of life that worked at forty-five but does not work any longer. The good ways to grow old all include a frame of mind that welcomes change and new beginnings and a talent for letting go. It was time for me to make changes while I was still young enough to make a new life, become part of a new community, make new friends in an easier place, where I could remain independent longer.

I had learned all that, but it was a matter of the head; there was still a problem with the heart. Although my farm was not the same place it had been before it was homesteaded more than one hundred years earlier, it was a very pretty spot, filled with wild things living out their lives in a way I hoped they found satisfying; I knew I did. My land was between a State of Missouri Natural Area and the first U.S. Park Service Scenic Riverway. The use restrictions on those pieces of public land extended the habitat I had

helped create. Twenty-five years of work had made the place a part
of me, even as I grew to understand that I, a particular human being, was irrelevant to it. In the end, it was that understanding from
the place itself, with its own species of magic, that taught me to let
go of what was not important—deed and title—and to hold to the
part of it that was—memories of my time there, the integrity of the
place, the lives being lived out there.

After we bought the house in Maine with its five acres, I got in
touch with the Missouri Department of Conservation, which
had indicated that it might be interested in adding my property to
the existing Natural Area. Natural Areas stay "forever wild." I
had learned at a very early age to be skeptical of the word "forever."
But if the MDC kept the place wild until governments fell, that
would be as good as humanly possible and the best I could do for
the wild things living there as I looked up the years to an unknow-
able future. The MDC people paid a visit, liked what they saw, and
by the time the demolition work in Maine was done, we had
worked out a deal.

The last week of September, on the day before I started pack-
ing up to leave Missouri, I had a party and invited everyone to
come. I was feeling sentimental and nostalgic and wanted to
gather around me people I'd come to love. I wanted to mark the
passing of my place from private to public ownership. The
weather cooperated; it was one of those clear, bright-skied autumn days. The Ozarks were at their winsome best. Goldenrod and
asters bloomed out in the field. Sumac leaves were bright red
around the woods' edge. Some of the best friends I've ever had
came to the party, good old boys, the reformulated hippies, the
town gentry, neighbors, and the feds from the Scenic Riverway.
I stopped talking with people at one point and took a mental
snapshot. Children, some of them honorary grandchildren, were
running around, tumbling, laughing, screaming. Couples were
walking out into the open field, heading back toward the woods

to walk the trails. My friend who makes home brew was dispensing it. Others were standing around eating from paper plates, talking, smiling. It was not a sad party. It was just the party I wanted, a cheerful one.

I'd asked the MDC to send a representative to mark the new ownership and, as a surprise to me, the young couple who would be moving into the house were the ones sent. He is a botanist, recently assigned to a new department position in the area. I took them around, showed them the house's quirks, pointed them in the direction of the trails. When they returned, they had fallen under the magic of the place and couldn't wait to move in. It turned out they even kept bees. If I'd written an order for successors, I couldn't have done better.

So in the very end, those ending days when I packed my belongings and drove down the driveway for the last time, I was content. And I took with me that photo album of the mind, images that could not be packed up and carried away by North American Van Lines. There was that one of the party, to be sure, and lots of others. Let me show you just one.

I am walking down to the mailbox, which is a mile and a quarter from the house. To get there I cut through the abandoned field belonging to a neighbor. The wind picks up, smells of weather, blows my hair across my face. The clouds, which had been promising rain, rush across the sky, lower, and turn a deep blue, slate. A flock of goldfinches, feeding on the seeds at the ends of grass stalks, startle, fly up. The grass turns emerald green in the strange light of the approaching storm. The finches are butter yellow against an iron blue sky. The beauty is mine to keep. The wind doesn't care, doesn't care, doesn't care.

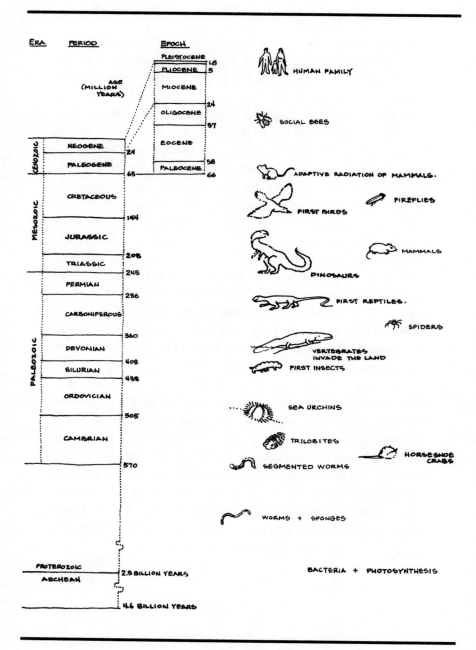

Time line of the animal kingdom.

1

I WENT TO THREE different colleges before I managed to snag an undergraduate degree, and considering how callow I was in those days, it is a wonder I learned anything at all. But looking back, I believe I learned three things. Two are irrelevant to our purposes here, but one has some bearing. A professor at the first of those colleges penetrated my attention sufficiently to impress upon me that there was no such thing as objective writing, that every inscription, every traveler's tale, every news account, every piece of technical writing, tells more about the author and his times than it does about the ostensible subject. The best that can be hoped is that the writer will lay out his bias up front.

I'm nearly half a century away from my college days, but everything I've ever read, including the scientific papers on biological subjects that I mine for my own writing, has convinced me that the professor was right. The writers of those papers, men and women I have interviewed and have come to know and admire, have selected their subjects because they have a passion for them. They have observed phenomena with an eye shaped by their experiences in particular places and times and have found interest and significance according to their own gifts, limitations, sadnesses, and sociabilities; their understandings have been shaped by their own peculiar and quirky worlds.

One of the better achievements of Western thought is the scientific method, which, when it works well, makes allowances for all those passions, limitations, and quirks. A good field biologist goes out and looks at the real world, sees something no one else has noticed, writes it down as accurately as he can, and reports it to others through publication. On the basis of his observation, he or someone else may spin a theory that explains it. Other observers with other passions, later observers with other biases, check it out, supplement, amplify, disagree, revise. Science is a process, not a body of received wisdom.

My own interests run to small animals that creep and jump and slither and flutter, the invertebrates — "the little things that run the world," as E. O. Wilson has said. What I hope to do when I write, keeping my old professor in mind, is to tell about the world of invertebrates, which appears to me an interesting and engaging one. I try to lay out my biases up front. I am grateful but astonished that over my writing life — one of the several lives I've had — sober, serious, responsible grownup editors have let me loose in that particular world to satisfy my own curiosity and amuse myself, in return for bringing back reports of what I see. There is a lot of news.

Because we have backbones — vertebrae — we think they are important, so in one of those smug Aristotelian bifurcations, we have divided up the animals into vertebrates and invertebrates. Actually, it was Lamarck, the eighteenth-century French naturalist, inventor of the notion that acquired characteristics could be inherited, who dreamed up the word "invertebrates." But the idea goes back to Aristotle, the original digital thinker, a man who always divided things into two categories. His basic division of animals was into those that had blood and those that lacked it. By blood he meant good, rich, red blood, not the pale ichor that oozes from a bug when you squash it, and in the blooded category he included mammals, birds, reptiles, and fishes — in short, vertebrates, animals with backbones. The bloodless were crustaceans, cephalopods, insects, and snails — in short, invertebrates, or at least the ones he knew about. His scheme made further divisions of the Twenty Questions sort. Does the blooded animal have hair or is it hairless? Legged or legless? If it has legs, are there four or two? This digital, yes or no, approach is familiar to anyone who has used a botanical key. Leaves alternate or opposite? Stems fuzzy or smooth?

Today, however, the basic principle of taxonomy is that living things should be defined by what they have, not what they lack, so taxonomic divisions come in varying numbers, not just in twos. The biological reference text I use divides all animals into thirty-two phylum units, of which animals with backbones are only one. But the Aristotelian way of thinking is powerful and still dominates our view of the world. College courses are taught and textbooks written on "Invertebrate Zoology." The premier zoological research institutions of the world, such as the Smithsonian, have departments of vertebrate and of invertebrate zoology.

Thoughtful taxonomists and zoologists grumble about this backbone fixation. One of them, Robert O'Hara, part philosopher, part zoologist, wondered how humans might be classified if an arthropod were making up the scheme. Arthropoda, meaning "joint-footed," is the biggest phylum of animals, those that by numerical rights, at least, should do the classifying if they were so inclined. The phylum includes spiders, insects, crabs, and lobsters. An arthropod might, according to O'Hara, describe us and our near relatives this way: "The anarthropods are a primitive group with few species and a limited diversity of form. Their reproductive rates prevent them from adapting to their environments closely, and the giantism exhibited by many anarthropods has kept their numbers very low and is no doubt the cause of their general sluggishness."

The invertebrates could get along without us quite nicely, and did for hundreds of millions of years, but we could not get along without them, so dependent are we on the life processes they have initiated and keep going. We humans are a minority of giants stumbling around in the world of little things, often not noticing our neighbors, not even being able to see many of them because they are *very* small. Yet each and every species, constituted from the same basic handful of chemicals as we are, has a complicated and special way of getting on in the world, different from ours and different one from another.

When we learn something about the way invertebrates live, they become familiar to us and we develop some charity and friendliness toward them. I am pleased to know that the woman who lives next door to Arne and me in Washington, D.C., has a home business that is thriving and that her children are doing well in school. I am pleased, also, to know that the bumblebee I see in early springtime, working the azaleas in our front yard, is a solitary mother queen hustling up enough provisions to raise daughters who will help her work during the months ahead. I find it satisfying and enjoyable to watch a periwinkle snuffling through the algae on a seaside rock near our new home in Maine and know that it is having a good feed.

Although we don't know everything about vertebrates, we know quite a lot: occasionally a new bird or fish or mammal or reptile is discovered, but those discoveries are rare, so we know pretty well how many tens of thousands of vertebrate species there are, and we know a good deal about their biology and behavior. But more than 95 percent of the animal species are invertebrates, and we have discovered only a fraction of the suspected tens of millions of them. We know next to nothing about how they get on in their lives or what happens when we lurch through their communities. We don't even have a fix on the number of invertebrates pecies already known and described, which is surprising in these days of computer technology and catalogs. Nowhere in the world is there any tidy master list of all discovered species with their names neatly registered and affixed. Specialists in a particular group — the fireflies, for instance — usually have a fair notion of how many species there are in that group, but even those are educated estimates, and the numbers within bigger groupings, such as insects or sponges, are guesses. So we don't even know what we know. Anyone venturing among invertebrates is sure to have the pleasure of discovery. And nearly every observation becomes a piece of news from an uncharted world.

That sort of pleasure is certainly one of the reasons zoologists go out into the field . . . and why I follow them there. In working with them I have been struck, as a writer, with the similarity of our occupations. Both journalism and field biology require standing aside a little, playing observer, not participant, taking a look at the real world, interviewing its inhabitants one way or another, watching the manner in which they get along, and reporting what we have seen or heard. But more than that, both writers and biologists share a compulsion to describe to others how we see that world. We keep at it, publishing this bit now, another later, piling up reports of our own realities.

A couple of books ago, I learned just how heady the pleasure of discovery could be and what a compulsion its pursuit becomes. In that book, *Broadsides from the Other Orders,* I was writing on several themes and using the biology and behavior of a particular bug to explore each one. I selected my bug examples in direct proportion to my own curiosity, which is to say ignorance, about them: silverfish, katydids, water striders, daddy longlegs, and so on. For twenty years I had been living, on my farm in Missouri, in close company with camel crickets — pale, silent, humpbacked crickets with stripy brown jumping legs. When I looked them up in guidebooks I found the same three or four tired facts about them, but nothing that satisfied, so I put camel crickets on my list for the book. My procedure, as chapter added to chapter, was to head down to the Library of Congress and spend several weeks reading everything I could find on the new chapter's bug. Usually, by the end of that time I had found the one person who was doing interesting work on it. I would telephone him or her and ask if I could come talk. Entomologists are generous people, and they invariably said yes. But with camel crickets the situation was different. Very little was known about them, and nearly all of it had been discovered by one man who had died by the time I started my reading. A student of his, on the verge of retirement himself, had continued some of the taxonomic work, and he kindly identified the species I had on my farm. He also informed me that my species was a special kind that had piqued his mentor's interest. In fact, he gave me a paper written by that man near the end of his life, in which he described the species as a piece of "unfinished business and a beckoning problem."

Briefly, the paper explained that my camel crickets existed in a small area in and around my farm as an isolated group within a larger population that were thought to be the same species. When the males of this group matured sexually, they became puzzlingly different from the males outside, whom they otherwise resembled. They grew a bright orange bump on the back of what we would like to call their necks but mustn't, because bugs don't have necks. And, the taxonomist told me, no one knew anything about how camel crickets mated, for that act had never been seen: perhaps the orange bump contained a food gift that males allowed the females to bite into during courtship. The contribution of a body part is a part of some insect wooings. Camel crickets of other species are common across the United States. They are nocturnal and live in basements, under piles of stacked lumber, in corners of garages, in country well houses, but none besides mine flaunted an orange bump.

I learned to sex even immatures, captured a few males and females, and began keeping them in a terrarium. I figured out

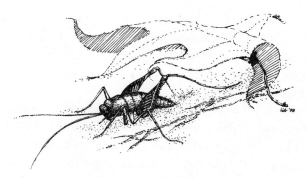

A mature male camel cricket (¾ lifesize). Note the
bump on his pronotum, just behind the head.

what they liked to eat, gave them water in bottle caps, learned that
they were virtually blind and deaf but received a startling array
of information through their more-than-body-length antennae. I
watched them molt and grow, and took notes on all this as much
for myself as for the chapter I was to write for the book. And, yes, I
became the first person to see them mate and lay eggs. They mated
in a contorted position, but the female never nibbled at the male's
orange bump.

For comparison, I began raising, in separate terraria, populations
of camel crickets from twenty miles south of my farm, outside the
range of my group. I discovered, to my surprise, that those males
and females matured more slowly than did the ones from my farm
and therefore mated much later in the season. This meant that in
the wild the two populations would never interbreed. Could they?
I isolated immature males and females from both populations and
kept them cloistered in single-gender terraria until the ones lacking the orange bump were ready to breed and then mixed them,
pair by pair, in still other terraria.

By this time my chapter on camel crickets had long since been
written and the book published, but I was still fired with the delight of investigation. In those days I was regularly commuting the

1,010 miles between my Missouri farm and the house in Washington in a van loaded with manuscripts, source books, two dogs, a cat, the unabridged dictionary, and the eleventh edition of the *Encyclopaedia Britannica*. To that pile I added a good many terraria of camel crickets.

In the end I found that the two populations could be induced, in this artificial way, to interbreed and that their offspring were fertile, so that by the common definition they did indeed belong to the same species. That definition holds that a species is made up of individuals that interbreed and produce fertile offspring. But the offspring of these forced matches were frequently deformed; some could not molt successfully and often died before their time. Does the orange bump of the speedily maturing group contain a hormone that hastens development? Are the two groups genetically distinct? Is the orange-bump group a species in formation or, alternatively, a primitive group? My farm bordered a geologically unusual area that contains plants relict from the Ice Age. Might my camel crickets be relict, too, from a time when the season was shorter?

I haven't the equipment, skills, or training to answer those questions, but they spurred me, when I finally admitted that I had to give up the farm, to place it with an organization that will keep it and the camel crickets forever wild and safe. There were, to be sure, additional reasons for that decision, but the presence of the unusual camel crickets contributed to it. It is my hope that someday a young entomologist will stumble across something I've written and go looking for camel crickets with bright orange bumps one springtime. He may answer those questions, but he will ask others. That's the way the process goes.

There will always be people who believe that watching camel crickets is an unsuitable way for a grownup to spend her time and energies. One reason for the lack of information about camel crickets was that although they are an extremely common insect, big and easy to observe, they have little relevance to humans; they neither help nor harm us. Most research is driven by the availability of funding, and funding agencies need to see a payback. Will research on this particular oddly named animal we have never heard of improve human life in some demonstrable way? If we fund the research, will we find a better way to kill the animal if we believe it to be harmful or to encourage it if we find it beneficial? Camel crickets, in that sense, are neither "good" nor "bad" bugs. In this they resemble most of the life on the planet. In the very short run, the one in which research funding takes place, it is hard to show that much of what lives and grows in the world relates to us. We eat salmon but not sea mice, the elusive *Aphrodite,* a genus of worms that live in the ocean deeps, where, some scientists say, there may be as many as ten million other species as unconcerned with our existence as we are ignorant of theirs. Certain black flies cause a terrible tropical disease in humans, river blindness, but camel crickets do not make us sick, nor do most of the more than one million other species of insects. Sensible, practical concerns should and do make salmon and black flies of more importance and interest in the funding of research and the formation of public environmental policy than camel crickets or *Aphrodite* or those other nameless millions. That is the way of the world, or at least of our time in the world.

I must confess, however, that I've never been convinced that to be interesting and important an animal needs to have a relationship to, or use for, human beings. As a matter of fact, otherness, remoteness, and independence engage my curiosity and intellect more than do similarity and utility. And I think there are a number of good long-term reasons why we should persist in our interest in camel crickets and sea mice and all those millions of invertebrates awaiting discovery. Some of those reasons will turn up in the pages that follow. But here, initially, I want to return to this compulsion that we have, writers and naturalists, to become tellers of tales about the world we have traveled in.

My brother, Bil Gilbert, a writer whose preoccupations run to animals with backbones, most especially crows, has often written of our "fascination with other bloods" as the basis of our yearning to understand the ways of others, as an escape from mere human perspective. But it is my son, Brian Hubbell, also a writer, a poet, who has said it best of all. Poets often do. He wrote, in a 1996 issue of the *Beloit Poetry Journal:*

The Message

Dry September weekday morning,
time indolence equates to sin,
I was outside, eyes closed over coffee
thinking some things I think when
a grasshopper flew in a certain fury
arcing off the desiccating asters
and affixed itself sharply to my upper lip,
returning startle with a compound stare
a ceaseless green on green.

Buttoned in its urgent grip
words came as plain as under
a falconed pigeon's ring
or in a half-corked bottle;
surely the heart of sadness is this knowing:
each embarks a gulf alone, each with
his fractured bit of seeing and a need
but not the tools to tell.

READINGS

Barnes, R. S. K., et al. *The Invertebrates.* Oxford: Blackwell Scientific Publications, 1993.
Brusca, Richard C., and Gary J. Brusca. *Invertebrates.* Sunderland, Mass.: Sinauer Associates, 1990.

Buchsbaum, Ralph, et al. *Animals Without Backbones*. Chicago: University of Chicago Press, 1987.

Clarkson, E. N. K. *Invertebrate Paleontology and Evolution*. London: Chapman & Hall, 1993.

O'Hara, Robert. "Telling the Tree: Narrative Representation and the Study of Evolutionary History." *Biology and Philosophy* 7, no. 2 (1992).

Ruppert, Edward E., and Robert D. Barnes. *Invertebrate Zoology*. New York: Harcourt Brace, 1994.

2

I FOUND THE BICYCLE at a yard sale. Its front wheel was a little out of round and the brakes didn't work, but it still had a couple of gears to choose from and it was a pretty thing, so I bought it. Back home I tinkered with it, and one morning I swung up onto the seat and pedaled to the oceanfront park a mile down the road from our new home in Maine. I wasn't allowed a bicycle when I was growing up because my mother thought they were dangerous. So of course I was mad to have one, and the first purchase I made at college when I was safely away from the parental eye was a secret bicycle, which I rode for years thereafter, feeling rebellious, exhilarated, and slightly wicked all the while.

The morning sky was deep blue, the sun shone through the trees lining the narrow road we share with the park. The air carried the fragrance of spruce and leaf mold I always associated with Maine during the years I came here as a mere visitor. Now I am a resident, and in the bicycle-created rush of air blowing my hair and jacket I felt, in my seventh decade, rebellious, exhilarated, and gloriously happy. The brakes still didn't work very well, but that simply added to the zestiness of the ride.

In the nineteenth and early twentieth centuries, before air conditioning, the well-to-do from lower latitudes, known as "rusticators," bought up the islands and promontories along the coast of Maine, including some on the peninsula where our place is. They built comfortable summer houses and called them cottages. The most beautiful spot on this peninsula was purchased by a Tammany politician who intended to build here but never got around to it. In the 1920s he donated the land to the town at the head of the peninsula. The park created from his land is a mere ten acres, but it is grand, if not in extent, in prospect, geology, and, as it turns out, intertidal invertebrate populations. I was bicycling down there, invertebrate survey census and field guides in hand, because I wanted to become acquainted with those populations.

I fastened my bicycle to a post around which a wild rose was blooming extravagantly. Beyond the roses, beyond the sheltering growth of fir, spruce, and birch, the rocks began: massive pink granite bluff and boulders with an exposure to the southeast. Waves crashed against the rocks and sent up wide plumes of spray, forced their way through fissures, where they became jets of water sparkling in the morning sunshine. To the south and east lay islands and, beyond them, the open sea. I looked outward until the earth's curve took away my vision.

The land here in coastal Maine is, geologically speaking, new and raw. Its creation and the extraordinary forces that accomplished it can be read upon its exposed rocks. The granite is sundered, its fissures filled with black basalt, making stripy roadways down to the sea. At ocean edge a rock outcropping shelters all behind it and helps create pools when the tide goes out. Beside the outcropping the boulders give way to heaps of smaller rocks and cobbles, testimony to transforming forces. Geologists can read these rocks and can read also the stories written by fossils of invertebrates that swam the ancient seas lapping far to the west of the Atlantic's edge today. Their readings change with new discoveries and revised geological theory, but at present the story is something like the following.

Formations of rocks along the Maine coast are the same as many in Ireland and England, telling of a time when the continents were configured differently. Seven hundred million years ago and for millions of years thereafter, an island that contained these formations, named Avalonia, lay to the east of the major portion of the North American landmass, separated from it by an ocean named Iapetos for one of the Titans, those beings who came before the gods. Avalonia included land that today can be found in Britain, Newfoundland, New Brunswick, Nova Scotia, Massachusetts, and Maine.

I have trouble understanding "700 million years ago," a trouble going back to my earliest years, when I spent more effort than was probably healthy struggling with adult constructs of time's extent. In those days my father was the superintendent of parks and cemeteries in Kalamazoo, Michigan. Part of his pay package was the use, for our home, of a fine old rambling house in one of the city cemeteries. Attached to the house, across a hallway, was the cemetery office, and the people who worked there were surprisingly tolerant of my coming to play in a corner of it, eavesdropping on cemetery business. One day I heard the secretary tell a customer that for the basic price of a burial plot the family would be responsible for mowing and maintaining the grave, but for somewhat more money, the cemetery offered "perpetual care."

What did perpetual care mean, I asked the secretary as soon as the customer left.

"It means we'll mow it, take care of it, forever," she explained.

"What is forever?" I asked.

"It means the grass gets mowed after those people die, after their children die, after I die, after you die."

Me? Die? That was new, too, and I wondered what happened next.

"Well, everything keeps on, you know. You'll have children and they'll have children . . . and . . . I think maybe it's time you ran back across the hall and went home."

For many years afterward, as I learned about things that lasted a long time — the British Empire, the Age of Dinosaurs, the formation of the solar system — I would check out the years of duration against this disturbing idea of perpetuity, of forever, and found that none of them measured up. I learned that empires fall, that animals become extinct, that the solar system had a birth and would have a death. I developed a strong suspicion that grass, maybe even trees, would grow over those cemetery plots someday as if they never had existed. Forever, it appeared, was a word made up by adults so they would not have to think about endings. Eventually, as was right and proper, I grew up and understood that perpetuity simply means until a government falls or loses interest. A friend who is an attorney told me not long ago that a recent national survey of legal documents shows that "forever" lasts about thirty years on average. But, if forever can mean until governments fall or lose interest, what does 700 million years mean when the whole history of governments, the very idea of governments, is subsumed into inconsequence by that span of time?

I have had more than twenty dogs as companions during my lifetime. None of those dogs has lived beyond sixteen years, but in that time has managed to pack in a full schedule of bumptious puppyhood, some good middle years of chasing rabbits and being a credit unto himself or herself, and sedate snoozes on a dog bed toward the end. My mother and father were both born in 1900; my father died young, but my mother lived well into her nineties, near the century's end. I can comprehend one hundred years without any trouble in thinking of their lifetimes, during which the telephone, the automobile, air travel, and the fax machine brought the world close; a time during which this country took part in several wars over matters thought important enough to kill for; a time during which we have rushed from premodern to postmodern. One thousand years? I can't comprehend that span in my bones, but intellectually I can take it in. It is approximately the time from the Norman Conquest of Britain to the present. It contains nearly all the kings and wars, philosophy, art, literature, music, scientific discoveries, and explorations that we call Western Civilization. One million years, a thousand of those thousands? That *is* hard. It is roughly half the time something designatably human has been around. But for the geologist or invertebrate zoologist, let alone an astronomer, it is simply a number tucked up into "margin of error."

"Seven hundred million years ago" becomes something on the order of "Long, long ago" or "Once upon a time," and that is not a bad way to tell of the beginnings of this beautiful little park, of the beginnings of Maine, although not of the beginnings of invertebrate life. For even then, even once upon a time, 700 million years ago, invertebrates had been around for a while, and life itself, cheerfully replicating and altering its forms, can be marked only in years numbering billions.

So, then, once upon a time, in an era called Precambrian, the ancient seas contained a rich worldwide population of soft-bodied invertebrate animals. Because their bodies were soft, they did not fossilize well. Also, many of the sediments into which they sank as they died are not accessible to us today because they have been subducted under the earth's crust by geological forces. But some fossils do exist from that time, among them those of sponges, worms, and cnidarians, ancestors of modern-day animals such as jellyfish and corals. Back before this park was formed, these animals were already complicated creatures with an evolutionary past.

During the millions of years that followed, Avalonia and the North American landmass spread apart and the ocean Iapetos expanded. As Precambrian times faded into what we call the Cambrian, the first part of the Paleozoic era, some 570 million years ago, the fossil record became better. It is, in fact, extraordinarily rich — the famous Burgess Shales date to Cambrian times — and includes a diversity of invertebrate forms that bewilder us today.

Sometime after that zestful Cambrian carouse with life forms, perhaps 400 million years ago, segmented worms of a sort crawled out of the safe, stable seas, with their equable temperatures and protective moisture, and took a chance on land. These were the ancestors of what we now call insects, and in a relatively short time, geologically speaking, they were flying and on their way to becoming the most successful class of animals the planet has ever seen. By Permian times, at the end of the Paleozoic, several hundreds of millions of years later, beetles, dragonflies, and a number of other familiar insects were already flying and scuttling about.

Five-inch-long trilobites, cousins of today's horseshoe crabs, fed in the sediments eroding into the ocean over what is now Maine and left their fossils there when they, along with the large majority of all living animals, were extinguished by the perilous geological, climatological, possibly even chemical, events that closed Permian times. But an array of invertebrates did survive, including not only insects but spiders, horseshoe crabs, true crabs, centipedes, and many, many worms.

During that time Avalonia and North America moved closer and gradually collided. The continent folded and buckled from the impact; the Appalachian Mountains were formed. The unimaginable pressure of the collision and the subduction of part of the land created earthquakes and volcanic eruptions. As the surface began to cool, huge masses of granite began to form under the shrinking sea that still covered much of Maine. But it would take the passing of many more years to create the conditions that forced the finely grained black basalt to ooze and intrude into the fissures of the crystalline granite, making those black bands within the pink rock at my park. It is hard to picture the so-solid black basalt liquid and hot, but the evidence was there beneath my feet that morning.

As Avalonia began to drift back eastward, it took with it and also left behind fossil-filled rock. The matching of the fossils in the rocks on both sides is part of the evidence that the coasts of Maine are of common origin with the coasts of Britain.

The evidential record of more recent geological eras is easier to read. Eighty million years ago a bee, very like the stingless bees still flying in Central America today, was flying in the forests that covered what we now call New Jersey, though it was much warmer in those days. The bee became entangled in a drop of sticky resin exuded by a conifer of a kind now extinct. She died in that resin, which in time became amber. In 1988 scientists figured out how long ago she had been entombed and declared her the oldest known bee. But, they cautioned, as a highly derived animal, she had an evolutionary past; primitive bees must have existed much earlier.

During the millions of years when the fossil record was being laid down, continents drifted and climates fluctuated, pruning out some species of animals, creating opportunities for others. About two and a half million years ago, as a period of volcanic eruptions ended, the climate began to cool and sheets of ice — glaciers — poured across northern areas of America and Eurasia. They battered their way across the land, scouring it, shaping it under three-mile-thick layers of ice and snow, creating here in Maine the land's profile we know today. The glaciers tied up so much water that ocean levels dropped hundreds of feet, and the sheer weight of the ice depressed the land.

During this time humankind may have been emerging. A mere eighteen thousand years ago, well within human times, the climate began to warm again and the glaciers began to melt, releasing water into the seas and releasing weight on the bedrock, which began to rise. The retreating glaciers left behind masses of boulders, smaller rocks, soil, and sand, filling in the profile of the land. Eleven thousand years ago the glaciers disappeared from Maine, but the sea level was still far below its present level here.

It was only five thousand years ago, as the Egyptian pyramids were beginning to be built, that sea levels along the Maine coast reached their approximate present-day levels, that this rocky strip of shore, the product of heat, uplift, and glacial rendering, began to look at all as it did the day I went looking for invertebrates.

I sat on one of those black basaltic intrusions, itself fragmented and cut, and looked out toward the sea, where the merest haze of fog was forming on the horizon. In back of me, delicate pink-blooming cranberries had invaded rock crevices. Wood lilies and beachhead iris bloomed against the fir trees. Nearby, a tangle of wild morning glories spread rock- and seaward. Within one pink-and-white glory, a half-inch-long wild black bee, her pollen baskets stuffed with white pollen, greedily packed in more. Zigzagging along in the sunshine at the rocks' forest edge were two sulfur butterflies, rich, sunny yellow.

The Register of Maine Critical Areas Program designates places in the state, some on public land, some private, that have rare natural importance. The invertebrate survey of the critical area that winds along the water's edge here lists only one insect — mosquito in larval form — because its focus is on the intertidal area. So I climbed down the granite boulders into the pools that the outgoing tide had left behind. The species survey list I carried was drawn up twenty years ago, when government spending levels matched environmental enthusiasm. A survey by the state of Maine declared this area and seventeen others along the coast, where the strip of land between the tides provides living space for a variety of invertebrates, worthy of being included in the register.

The survey found thirty-four species of marine invertebrates here, some of them small, but all visible to the naked eye. They had taken up life in one of the most testing environments the planet has to offer: exposed to harsh, drying air at low tide, vulnerable to any sharp-eyed sea gull; covered with salt water but tugged by waves and currents at high tide, and prey to a whole range of ocean animals looking for dinner. Those very stresses have promoted innovative adaptations. Some animals are attached to the bottom and appear plantlike. Others swim or crawl. Some are primitive, others are very specialized and highly adapted. Some have shells or other protective gear. Some have soft bodies. The presence of one common soft-bodied invertebrate infrequently found here, a species of nudibranch (the name means naked gills), *Dendronotus frondosus,* is one of the reasons this strip of tide-washed rocky shore has been declared a critical area. I have never seen *D. frondosus.* I had never seen *any* nudibranch that I am aware of.

Peter F. Larsen, a marine invertebrate zoologist at the Bigelow Laboratory for Ocean Sciences in West Boothbay Harbor, was the principal investigator on the invertebrate census done here. I asked him why a seldom-seen animal such as a nudibranch, of no apparent human utility, was important enough to have earned this spot the designation as a critical area.

"Well," he said, "the nudibranch may not sound important. But it is rare and who knows? Fluctuations in a population of an unusual animal like the nudibranch can alert us when an entire system is in trouble. These marine systems are complex. We don't have any idea how many species there are in the ocean, but we do know that they represent a greater biodiversity at a higher taxonomic level than on land. And we need all of the biological diver-

Dendronotus frondosus (lifesize).

sity that there is to maintain the functioning of the entire ecological system. When a system becomes impoverished in species, it collapses rapidly." He paused and then added, with a laugh, almost apologetically, "And besides, a nudibranch is one of God's creatures too, and it is interesting all by itself."

Hank Tyler, now a State of Maine planner, was administrator of the Critical Areas Program until funding was cut off in the 1980s. He is a rare combination of passionate zoologist and enthusiastic bureaucrat, and one afternoon he outlined a history of the program, pointing with pride to the animal and plant census created, the pile of publications the program had generated, and the heightened public awareness of rare natural communities the program had helped create. I asked him the same question I'd asked Peter Larsen: why should we take an interest in a rare animal with a funny name like a nudibranch?

He paused to reflect on my question. His answer, when it came, was a quiet and personal one. He said, "I think we are attracted by beauty and rarity. We like unusual things — a precious stone, a diamond, an original painting, a rare bird, and those animals with strange names when we find them. And somehow . . . I don't quite understand this . . . maybe it is innate . . . once we know about them, we want to protect them."

Nudibranchs, often called sea slugs, are rather like snails without shells and belong to the same taxonomic grouping, the gastropods. The *D. frondosus,* reddish brown with white spots, is a couple of inches in length, tapered at both ends, and covered with a forest of fancily branched gills through which it breathes. It is a good swimmer. It is found in the company of other small invertebrates upon which it feeds, hydroids, small, lacy, plantlike animals that defend themselves by stinging. The favorite hydroid of *D. frondosus* is reported to be one called *Tubularia,* which grows in dense pinkish masses on pilings, seaweeds, and other underwater holds. *D. frondosus* uses its jaws to tear out chunks of *Tubularia* but cleverly avoids being stung in the process; the stinging cells pass through its digestive system to the ends of its bushy gills, where they give the naked animal protection from predators who might fancy nudibranch for dinner.

The tidal pools bloomed with strong color: the coppery browns and greens of Irish moss and rockweed, the delicate greens of sea lettuce, sponges, mosses, and algae; the oranges, yellows, and reds of other algae. The sun sent glimmer and shine into the ripples in the small pools, ranging in size from soup bowl to swimming hole.

Barnacles, one of the thirty-four invertebrates listed in the census, are clubby crustaceans, kin to shrimp and lobsters. They have taken up adult life in crowded numbers here, glued to the rock surfaces among the weeds. Louis Agassiz, the eminent nineteenth-century founder of Harvard's Museum of Comparative Zoology, once described the barnacle as a "shrimp-like animal standing on its head in a limestone house kicking food into its mouth." As the tide recedes, exposing them to air, barnacles close up shop, withdraw the feathery feeding parts (with which they kick food into their mouths) into their hard, shell-like plates, and sit out the dry spell. Massing on the wet and often slippery rocks, they become,

A tide pool at my park, with the ocean beyond.

willy-nilly, nonskid pavement for humans. My list tells me that the barnacles here belong to the species *Balanus balanoides,* a Greek name that means "acornish acorn."

Fastened to the rock surface by a glue of their own making, these barnacles appear to lead a conservative life, growing by adding calcified material along the edges of their plates and waiting for the floating kind of dinner they can grasp and pull in. But their lives have had a more active phase. In autumn, a barnacle, with enormous effort, extends his male appendage and deposits sperm into his neighbor's shell. The eggs incubate over the winter and hatch out into bristly, triangular, free-swimming larvae that float along with all the other larval and plant life and bits and pieces we call plankton, a nutritious suspension for many animals. In the plankton the young barnacles feed or, more often, are fed upon. But if an infant barnacle is lucky enough not to be eaten during its six growth stages, it leaves the floating life and begins to search out a place to call home, using chemical cues to settle with others of its species. It uses its swimming legs for the last time to kick away the silt in a suitable spot, where it cements itself into place and begins to surround itself with six separate overlapping shelly plates, beginning the cycle all over again.

On the outskirts of the barnacle village, but sometimes invading it, are blue mussels, *Mytilus edulis,* the same that adorn plates of pasta and tomato sauce in Italian restaurants. These, too, spend the first part of their lives drifting in the plankton and, when mature, settle down among their own species, attaching themselves to a secure surface by a thread of their own spinning. Once attached, they filter out what is good to eat when the seawater covers them and close up their dark blue-black oval shells when the tide goes out. Underwater they spend a good deal of time grooming their shells, using the licking motion of their tonguelike foot to do so. That helps prevent other mussels from attaching to and smothering them. They are never completely successful, however, because great wads of mussels, fastened to one another, are all over the tide pools I waded in and anywhere else mussels are found. Thickly populated communities of living things, plant or animal, are always an invitation to dinner for other animals, and these mussels and barnacles are no exception. Sea gulls and people prey upon mussels exposed at low tide. When mussels are covered by water, starfish use their strong feet to pry open the shells and get to the tender meat inside.

Barnacles are often smothered, victims of the piling up of their own kind or of aggressively expanding populations of mussels or seaweeds. Like the mussels, they are also eaten by predatory snails. Snails look so unthreatening to us that it is hard to imagine what a single-minded and scary enemy they can be to a little guy glued into place with just a shell for protection. One of the commonest snails I saw in the tide pools that morning was the carnivorous Atlantic dog whelk, *Thais lapillus*. It attacks barnacles by secreting a poisonous purple dye (prized by the Phoenicians to color clothing) over them, then inserting its feeding parts into the relaxed plates.

With mussels, the whelk's strategy is different. With a tonguelike structure called a radula, it drills a hole into the mussel's shell, through which it inserts its feeding parts. The barnacles are defenseless against whelks, but mussels fight back. Sensing the snail, the mussel sometimes ensnares it with the thread it uses as a holdfast; what is more, neighboring mussels sometimes join in the attack, netting the whelk from all sides and flipping it over. Exposed in that position to predators or to drying air as the tide goes out, the whelk dies. One authority estimates that 30 percent of Atlantic dog whelks in mussel beds are trapped this way. The color of the whelk's shell depends upon what it has been eating. That day I saw both whitish ones, which had been dining on barnacles, and darker-colored ones, which had concentrated on mussels.

As the tide continued to ebb, I saw other kinds of snails, periwinkles, moving about with what seemed to be great rapidity for an animal we think of as the synonym for "slow." While I crouched over the pools contemplating dinner and death among the mussels and barnacles, my peripheral vision kept recording the hunching along of periwinkles. The inch-long, drab periwinkle snail, *Littorina littorea,* which becomes such a tasty morsel in French cookery, has a radula too, but instead of using it to do in the other animals of this intertidal village, the vegetarian periwinkle uses its tool-tongue to scrape algae from rocks. Scraping and eating. Scraping and eating. Scraping and eating so effectively that periwinkles can erode rock and alter the entire community by preventing seaweed from attaching. Periwinkles are tough. They have extra blood vessels that allow them to breathe air when they find themselves above the receding tide. When one is exposed, it fastens itself to a rock with a gluey mucus that hardens, sealing the vulnerable animal inside so that it will not dry out. On bare rock, if not picked off by a sea gull, a periwinkle can withstand temperatures as high as 111 degrees Fahrenheit. As the heat rises, the periwinkle enters a reversible state of heat coma and shuts down body processes. In the winter it can tolerate temperatures down to at least 10 degrees F by allowing ice crystals to form between, not in, its cells.

Impregnated females release egg capsules into the receding tides, where they join the plankton. As the eggs drift along in the southerly currents, they hatch into free-swimming larvae and, when they are mature, come in to shore far south of where their parents lived.

Periwinkles are considered to be European natives, foreigners to these shores, for they were first noted along the coast of Nova Scotia in 1840 (although there is a puzzling find of their shells in a Nova Scotian Indian midden that carbon-dates to A.D. 1000). They rapidly spread down the coast, displacing native snails in the process.

Atlantic dog whelks are sometimes called winkles, too. This splendid word is the occasion of rhyme for Ralph A. Lewin, a distinguished British-born phycologist, or algae and seaweed specialist. The following is from his 1987 collection, *The Biology of Algae and Diverse Other Verses.*

Song of the Winkles

I've seen the merry Irish on Saint Patrick's Day parades,
And throngs of sorry sophomores a-waiting for their grades;
I've seen a herd of buffalo a-chewing of the cud —
But I've never seen the wily winkles schooling in the mud . . .
Winkles, winkles, wily winkles, schooling in the mud.

I've seen the fans a-surging and converging on the gate,
And gangs of geese a-gathering, preparing to migrate;
I've seen the Holy Rollers rolling wholly on the floor —
I've never seen the winkles in the wrinkles of the shore . . .
Winkles, winkles, wily winkles, schooling on the shore.

I've seen a lot of lemmings milling madly to their graves,
And porpoises a-spouting on an outing in the waves;
I've seen a troop of hooligans a-fooling in a band —
But I've never seen a single winkle schooling in the sand . . .
Winkles, winkles, not a winkle schooling in the sand.

The ways of little animals are like the ways of men:
They go off in one direction, then come trooping back again.
The summer's nearly over, boys, so what're we waiting for?
Let us join the wily winkles and go schooling on the shore . . .
Winkles, winkles, wily winkles, schooling on the shore.

With my invertebrate survey list and field guides, I had been able to sort out some of the seaweeds and identify a number of invertebrates on the list, but like Lewin with his wily winkles, I had not found the rare nudibranch. That is all right; the quest is the part that is fun. And it had been a very good morning. I stood up to stretch; the fog, originally just a cottony line on the horizon, had closed in over the islands, although the sun was still shining brilliantly where I was standing. A tall, rangy young man carrying a plastic bucket was making his way across the rocks toward me. He told me he'd been gathering snails to steam. "Ever see a sea cucumber?" he asked. No, I hadn't, but I'd heard of them. There was even a "pickle plant" nearby, a sea cucumber–processing factory that bought them. In the pictures I had seen, they appeared to be big, globby-looking animals with feathery tentacles who moved around on small tube feet. I was surprised to find that they belonged to the Echinodermata, a phylum that includes starfish.

Lying limply in the young man's bucket of snails, the sea cucumber was white with blotches of brown and pink, about eight inches long. It was covered with warty-looking bumps and did indeed resemble some peculiar, bleached-out cucumber. I picked it up. It felt like a rather soft, flaccid pickle. It did not seem happy. It had re-

Cucumaria frondosa (⅓ lifesize), a sea cucumber common in Maine waters. Its tentacles are retracted.

tracted its tentacles and was smooth at both ends. "Now if you were to take it and throw it out in that pool," the young man said, "it would put out little feelers. D'you want it?" Indeed I did. "Don't you find it slimy?" he asked. I was supporting the poor animal with both hands. It felt cool and leathery and limp, a little like a damp, deflated football, but it wasn't slimy.

The young man walked on, and I went down to the deepest tide pool and eased the sea cucumber into it. I wondered if it had died. But after a few minutes, during which its blotches rearranged themselves, it released its tentacles from its mouth and plumped itself up. It seemed content in the water, so I left it, waded back out, and climbed up toward the boulders.

The receding water had left exposed any number of leggy overhangs, and under the first I came to I could see starfish clinging to the rock. Zoologists have declared that because they are not fish, we should call them "sea stars." They are spiny, five-legged cousins not only to the sea cucumber I'd just released but also to sea urchins and sand dollars. These particular ones, listed on my census sheet as *Asterias vulgaris,* the common sea star, are perhaps six inches across, or maybe a little more. Against the gray rock, their garish, tarty color catches the eye: brilliant pink-purple with orange and yellow highlights. I saw one, then another, and began to count: four . . . six . . . nine . . . I gave up because they were everywhere, resting out this low tide upside down, clinging to whatever rocky projection they could grasp with their muscular suction-cup feet. Their stomachs were probably full of mussel meals. I stopped to right a big orange-red rock crab, another animal on the list, *Cancer irroratus,* that the tide had left on its back. His color echoed that of the sea stars, on whom *it* may have been feeding. It had a big wedge-shaped carapace, maybe four inches at its greatest width, and powerful claws. It could have given me a pinch with them, but didn't.

Overhead, gulls were screaming. They wanted me gone so they

could feed in privacy on what the tide had uncovered. Perched on a fir where the woods began, a raven commented loudly on my presence.

The fog had obliterated even the near rock outcroppings by the time I reached their summit. The air had turned to pearl. I bicycled back home in its sheen and glimmer and remembered that these misty shores could just as well be England.

Back home I looked up the sea cucumber and found that the particular one I'd been holding was the orange-footed sea cucumber, *Cucumaria frondosa,* and that it was at home in rocky pools along these shores. More than a year later I liberated a pair of them from their trip to the pickle factory so that a picture could be drawn of one of them for this book. While they sat for their portrait, they characteristically altered their shape and color. Afterward we put them back in the ocean.

READINGS

Bachand, Robert. *Coastal Atlantic Sea Creatures.* Norwalk, Conn.: Connecticut Maritime Center, 1994.

Bennett, Dean B. *Maine's Natural Heritage.* Camden, Me.: Downeast Books, 1988.

Coulombe, Deborah. *Seaside Naturalist.* New York: Simon & Schuster, 1992.

Doggett, Lee F., et al. *Intertidal Bedrock Areas of High Species Diversity in Maine.* Bigelow Laboratory Technical Report no. 1–78, Critical Areas Planning Report 55. Augusta, Me.: State Planning Office, 1978.

Gosner, Kenneth L. *Peterson Field Guide to the Atlantic Seashore.* Boston: Houghton Mifflin, 1978.

Meinkoth, Norman A. *National Audubon Society Field Guide to North American Seashore Creatures.* New York: Alfred A. Knopf, 1995.

Stanley, Steven M. *Earth and Life Through Time.* New York: Freeman, 1987.

3

ONE DAY several years ago I received in the mail a fat envelope from Gary Raham, a biologist, educator, writer, and illustrator. In the envelope was a letter and an article he had written for *American Biology Teacher* entitled "Pill Bug Biology." He recommended pill bugs to me, said he thought I would find them interesting.

I did.

Pill bugs are those little creatures you find under a rock or in some other protected place. They curl up when you touch them. After I read his paper, I started turning over rocks and piles of old boards quite regularly to look for them. When I found some, I'd take them up in my hand to study. They are oval-shaped, flattened animals, looking a little like miniature tanks, perhaps half an inch in length. The ones I found were usually dirty gray in color, and if I could keep them from curling up defensively I was able, with a hand lens, to count their very short legs — fourteen in all, growing out of seven distinct body segments. This means they are not spiders, which have two-part bodies and eight legs, nor are they insects, with three-part bodies and six legs. Pill bugs are something quite different: they are a particular kind of isopod crustacean, most of which are marine animals, including the crabs and barnacles I watched at my park as well as lobsters and shrimp. *Isos* means equal (as in isosceles triangle) and *pod* means foot. Indeed, the pill bugs' fourteen feet, and the legs attached to them, are all about the same size. I could also see their pair of antennae and even their eyes, behind the antennae on either side.

When I put Gary's paper in my files, I discovered that I had a scientific paper on terrestrial isopods in Michigan, written by my old friend Arlan Edgar, who otherwise specializes in the biology of daddy longlegs. I reread his paper and was hungry for more, so since that time I've kept my eye out for more information about pill bugs and have developed a habit of poking in the leaf litter and under the edges of things to look for them.

These appealing little animals like nothing better than to nuzzle up against one another in a companionable group. They do a nice job for us in processing dead leaves and waste material, turning them into soil. Pill bugs harm us not at all; they cause no disease to us or our domestic animals. They are, as Stephen Sutton, a British zoologist from the University of Leeds, writes, "amiable companions to mankind." Quite a lot is known about them, particularly in England, where they appear to be fewer in number than they were one hundred years ago. Sutton offers a possible explanation. "Most of the older records were made at the turn of the century,

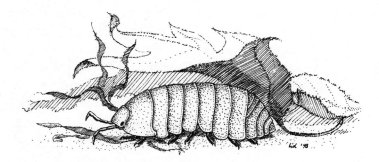

A pill bug (4 × lifesize) walking toward cover.

when hunting them enjoyed a considerable vogue among the clergy. Undoubtedly the low level of interest in the group is partly responsible for the scarcity of recent records, but misidentification and a previous tendency to do one's collecting in the rectory greenhouse may also be a part of the explanation."

Pill bugs are members of a land-living isopod family called, naturally enough, Armadillidae, for they can curl up as armadillos do and they present a somewhat similar platelike surface. Curled up, with their antennae retracted into grooves, they become little balls. Spiders who would like to eat them find no purchase on the round surface, and even larger predators, such as shrews, are reduced to pushing them around, ineffectually, in circles with their mouths. Some other isopods live on land, too, and look rather like pill bugs when they are flat, but the others lack the happy ability to roll up in a ball when disturbed. Collectively, the terrestrial isopods are commonly called wood lice. Some few isopods who made it to land went back into the water in fresh ponds and lakes, but most remain in the sea, where they flourish along with the 40,000 other species of crustaceans. In fact so few are the landlubbers and so recent is their arrival that they seem something of an oddity, reckless explorers that are not quite at home here yet.

The earliest fossils of wood lice are found in Baltic amber that dates to no more than 35 or 40 million years ago. Those creatures were highly evolved, similar to modern forms, and they were widely distributed, so obviously some ancestral forms had crept up onto the shore long before that. The usual opinion about wood lice is that they have not yet had time to make adaptive physical changes from an environment of water. The story they tell is one of the evolution of behavior more than of bodily form.

It is true that ancestral isopods still living in the sea had taken certain bodily forms that made invasion of the land easier for their descendants than it would have been for, say, a shrimp. Somewhere back in the richness of time, isopods developed to take advantage of ocean bottoms, where all the tiny edible bits settled, bits ignored by animals that lived and swam in higher strata of the sea. They developed short, stout legs to walk on the bottom and a flattened shape that withstands water pressure and as a result is very stable. Their close relatives the amphipods, which include the sand hoppers, or beach fleas, are compressed laterally — that is, they are taller than they are wide — and have never become very successful on land; perhaps, scientists speculate, they may be unstable.

By lucky chance, or by adaptation that increased the survival of their young, isopods while still living in the sea developed a kind of reproduction that became helpful to those who left it. Some land animals must return to the sea to reproduce, but male isopods have abdominal legs that are modified to transfer sperm directly into the females. And the females keep their young in a water-filled pocket, rather like a kangaroo's pouch, that keeps them nicely moist and protected until they are able to look out for themselves. As a consequence very few young pill bugs, or their watery kin, for that matter, die in immature stages.

Those were lucky cards to be dealt for life on land, but they were the only advantages the wood lice had. We are reminded that life originated in the sea by the fact that 70 percent to 90 percent of all living tissue is water and that the chemical reactions necessary to life all require the presence of water. Animals that live in water can replace their internal supply easily, but an animal on land must work out how to prevent its internal fluids from leaking out and how to replace those that do. Insects, spiders, and their kin, which have been on land for hundreds of millions of years, have developed a hard, waxy cuticle to keep from drying out; also their respiratory system of tracheae, finely branched tubules, bring oxygen to every part of their bodies. The tracheae can be closed by valves when necessary to prevent the loss of moisture.

Some species of pill bugs have a cuticle that gives them some small protection against drying, but they are leaky little animals compared to insects and will dry out and die when the humidity in the air enveloping them dips below 95 percent. Most still breathe with gills only slightly modified from those of their water-dwelling cousins, and even those that have developed rudimentary tracheae lack valve-controlled openings.

At one time some biologists considered the permeable cuticle of pill bugs to be a disadvantage for terrestrial life, a shortcoming that adaptation in some hundreds of millions of years would eventually correct. That may be true, but we should be cautious lest our own biological bias skew our interpretation. Stephen Sutton, the zoologist quoted above, has written an affectionate, provocative monograph called, simply, *Woodlice*. He points out that having a leaky body covering actually may be efficient in terms of energy expenditure. There is a long-held biological principle that the nitrogenous waste products of metabolism, ammonia in particular, are toxic to animal tissue and must be excreted speedily, as in urine, or detoxified by being concentrated in urea or uric acid. But, he cautions, invertebrates, particularly the higher crustaceans, which is what isopods are, can tolerate levels of ammonia in their blood that would kill a vertebrate. The wood lice produce no urine at all but release ammonia as a gas that diffuses through the cuticle. "This," writes Sutton, "opens up the intriguing idea that woodlice have a permeable cuticle not because they failed to come up with anything better in the course of evolution, but because the loss of water is less of a drawback than the loss of energy involved in detoxifying waste products."

They have evolved behavior that saves water in other ways, too. And Sutton's "intriguing" notion is a good one to keep in mind while considering the efficient ways pill bugs comport themselves to keep pleasantly damp but not too damp — efficient enough so that some species live in deserts and other dry places.

The isopods had short sturdy legs when they came onto land,

legs adapted to walking along ocean bottoms, and on land they became very good and speedy walkers. That is obvious to anyone
who has lifted up a rock and watched them scurry away looking
for other shelter. Rocks and old boards trap damp air. So the first
useful behavioral rule is "Walk toward shelter."

Indeed, nearly all their adaptive behaviors involve walking
briskly along on those fourteen stout legs. Elegant experiments
have revealed what pill bugs walk toward and what they walk away
from. Not only do they walk toward shelter and away from the dry
open air, they also walk toward cooler places, away from hotter
ones. So behavior number two is "Walk toward cool." Those
cooler places are damper (because cool trapped air can hold more
moisture than warm air), and the pill bugs can take up water
through mouth or anus when they need it, or they can eat damp
soil to extract moisture.

Many species walk toward dark and away from light. They even
prefer to walk on dark colored surfaces rather than light ones.
These preferences combine to turn them into nighttime feeders,
going abroad when the drying sun has gone down. Behavior number three is "Walk out when it is nice and dark." And the fourth is
"Walk to still places." Many species walk away from wind, which
removes the shell of damper air around them. If a breeze comes up,
they will seek a protected spot, even when foraging at night.

During extremely hostile conditions, a pill bug can keep to its
burrow and eat its own dung, recycling nutrition and moisture.
Even in the best of times pill bugs recycle their waste to reclaim
the copper lost in excretion. Their blood uses copper, not iron, as
ours does, to carry oxygen, and a pill bug prevented from eating
its own droppings will, in fact, die from lack of copper. On their
nighttime foraging expeditions they eat a lot of decaying plants,
small pieces of fruit, fungi, the droppings of other animals, insect
eggs. They eat nearly anything that is small and doesn't move.
Zoologists call such animals detritivores. And although they may

not be in the same recycling league as earthworms, they do help process this organic material and enrich the soil.

Pill bugs like to walk toward objects until they can touch them all over and then they stop. Zoologists have coined a twenty-five-dollar word for this: thigmokinesis. It describes a behavior that pill bugs share with some other residents of the soil and leaf litter. They keep moving as long as their feet touch things. When other parts of their bodies begin to touch something, they slow down, and when they get really cozy they stop moving. Because they all behave the same way and because they are chemically sensitive to their own kind, the "things" they are touching are often one another. They congregate, sharing the dampness of group respiration, in a big friendly crowd, in cracks, under stones or lumber piles, between blades of grass. Behavior number five, then, is "Walk to keep in touch."

Altogether, these behaviors help them conserve moisture. However, any animal that has truly adapted to land has lost its ability to live *in* water. Pill bugs can drown in a drop of it. So when their bodies contain too much water, their behavior reverses and they walk toward dry places. However, this ability to change behavior depending upon internal conditions and needs may be their undoing when it is slightly distorted by a clever parasite that specializes in behavior modification.

Parasitism is a kind of symbiosis. Symbiosis, when biologists use the term, simply describes close relationships that organisms of one species have with those of other species. It is not always a happy relationship for both parties. When it is, when both sides benefit, the symbiosis is called mutualistic. Sometimes, however, only one creature benefits and the other loses. Then the symbiosis is a parasitic one.

Acanthocephalans are members of a phylum of parasitic worms of some 1,150 species. Their name, which is descriptive, means spiny-headed. All acanthocephalans have two life stages and must live within two quite different kinds of animals. In juvenile form they live within arthropods. At least one species has followed the crustacean isopods to land and now lives within our pill bugs. But the adult reproductive form must live inside a vertebrate. The species that infest ocean crustaceans move on to fishes; the kind that infests pill bugs, *Plagiorhynchus cylindraceus,* must be taken up by songbirds, such as starlings, in order to live on and reproduce.

We think of parasites as loathsome, lacking somehow in moral fiber, and we have a hard time appreciating the pruning, limiting, winnowing effect they have on populations, allowing only the hardiest to survive to pass on their genes. They have a place in the adaptive process, yet we do not like them. I was reminded of this not long ago when I was reading about passenger pigeons, birds that once filled the American skies and were hunted to extinction. At the time, because they were so numerous and produced so many droppings wherever they congregated, they were regarded with the same lack of affection as city pigeons are today, but now they have been extinct long enough to become symbols of man's destructiveness. The last passenger pigeon was named Martha; she was stuffed and put on display in the Smithsonian, but the lugubrious placard that accompanies her does not mention that with her

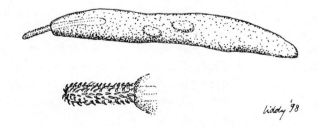

Top: The adult form of *Plagiorhynchus cylindraceus* (6 × lifesize), the acanthocephalan parasite of pill bugs and starlings.
Bottom: Its rasp proboscis, enlarged.

death two species of lice that had specialized in passenger pigeons also became extinct.

If *P. cylindraceus* doesn't measure up to our human ethical standards, at least we can credit the ingenuity of the adaptive process that has made it what it is. Pill bugs, as noted above, eat their own dung as well as that of other animals: gut analysis has shown that nearly 10 percent of their food consists of dung, including bird droppings. The pill bug, then, has only its own tastes to blame for picking up *P. cylindraceus* larvae from the droppings of infested starlings. After a larva emerges within the pill bug gut, it uses its spiny proboscis to rasp through the gut wall and drops into the pill bug's body cavity. Within two months it has grown to fill nearly half of the pill bug's body and is ready to be taken up by a starling. When a bird eats the pill bug, *P. cylindraceus* sinks its prickles into the bird's intestine. The parasite is so bathed in nutrients there that it needs no digestive system of its own; adult forms of *P. cylindraceus* are essentially reproductive sacs, male and female, on auto-feed. When fertilized, the female's ovary sloughs off tiny encapsulated larvae that pass out with the starling's droppings.

When Janice Moore, an English major turned parasitologist, began her work on *P. cylindraceus,* she found that pill bugs generally make up a very minor part of a starling's diet. The reason for this is obvious: pill bugs behave in ways that keep them from being eaten by birds very often. They usually go abroad in the damp cool of night, and if they are out in the daylight they walk on dark, camouflaging surfaces. It may not be that starlings don't find pill bugs tasty, it may be simply that they seldom find them.

Moore also discovered something that might appear puzzling at first: in areas where the parasite was present in significant numbers, *P. cylindraceus* was found in more than 40 percent of the starlings but in fewer than 1 percent of the pill bugs. She knew that acanthocephalans had a long history of parasitizing arthropods and that they had become specialists in behavior modification. So she wondered if perhaps *P. cylindraceus* modified pill-bug behavior in some way that would make them more likely to be eaten by starlings.

In her laboratory she fed some clean pill bugs bits of carrot with the worms' encapsulated larvae on them and fed a control group carrots that were not laced with larvae. When she put the two groups together, she could not tell from their appearance which pill bugs were infested with parasites and which were not. But she found that some had turned into risk-takers: they walked away from moisture toward dry places; they would just as soon be exposed as under cover; and they seemed to prefer light backgrounds to dark ones. When she sacrificed them, she discovered that the risk-takers, otherwise completely indistinguishable from the conservatives, were the ones that were hosts to *P. cylindraceus*. Moving back out in the field she found, through experimentation, that, yes, although only a minority of the total pill bug population in the wild was infested with the worms, that minority was most likely to be eaten by starlings and fed to their nestlings. Those pill bugs were candidates for starling dinner because they were in places starlings could see and catch them. The more conservative, worm-free majority were not.

In some ways this might seem to be quantifying the obvious: diminished, impaired animals are likely to come to bad ends. But it is a little more complicated than that, and you can almost hear the drumbeat of evolution while thinking about it. It is suicidal for a parasite with a single host (as some parasites have) to be so successful at its business that it kills that host before the parasite is able to reproduce. Perhaps some greedy parasites have done just that, but their genes have not been passed on, and no one lived to tell the tale. In order to survive, parasites of a single host often become self-restrained and courteous. Night-flying moths, for instance, have ears that are keen enough to hear the high-pitched squeaks made by bats in echolocation, squeaks too high for our ears to hear, and can evade the bats that hunt them. A certain mite that infests those mites lays its eggs in the moths' ears, which disrupts their hearing. However, by laying down a chemical trail as they climb aboard, the moths lay their eggs in only one ear. They spare the other ear, lest they deafen the moth completely and become, along with the moth, bat dinner, which they don't want to be.

That is one strategy of parasitism: be careful and preserve the resource. But for parasites that do get eaten, there is another strategy: adapt and don't let your eggs get digested. Selective pressure favors parasites that have such a makeup, for they will pass on the genes that allow such a happy outcome — happy at least from the parasite's point of view. This is the result we find in the life cycle of *P. cylindraceus* within pill bugs and starlings. When such a relationship has evolved, the first host (in this case, the pill bug) becomes expendable, but the secondary one (the starling) must be more cherished. The starlings suffer very little from being infested by *P. cylindraceus.* They may lose a little weight, but the bad effects usually are not serious. And the extra nutrition that pill bugs, even infested ones, provide may make it worthwhile for the starlings to eat them. For pill bugs the infestation is a disaster; for starters, they usually get eaten, and in addition, female pill bugs that harbor the worm do not develop ovaries. That is okay from the worm's point of view, for the pill bug is just a disposable. The starling, however, is the habitat in which the parasite reproduces. To kill the bird would be to cut short its own reproduction.

How does *P. cylindraceus,* a gutless worm, manage to do something as complicated as altering every survival behavior that pillbug evolution has brought about? In a 1984 paper, Janice Moore wrote that the alteration might be accomplished biochemically.

I wondered if she'd found out anything more about the way *P. cylindraceus* brings about the self-destructive behavior in pill bugs, so I telephoned her at her office at Colorado State University, where she now works. "Well," she said, "I've not worked any more on pill bugs — I'm investigating other animals now — but I'm still interested in parasites that change a host's behavior. I've just recently published a paper on the subject. I'll send it to you. And, yes, at least in some other animals with acanthocephalans, the change is probably biochemical. We're beginning to think that serotonin may have something to do with it."

Her new paper summarizes many examples of parasites altering the behavior of their hosts. In most cases behavioral change means the parasite is more likely to be taken up by the other host and thus be assured of a future. She ends the paper by asking a lot of questions. "What are the switches that some parasites have found to reverse responses to a simple stimulus such as light?" and "If parasitized animals forage differently, how is nutrient flow affected?" and even more broadly "How does the presence or absence of a parasite affect community structure?" But her most interesting question is this one: "When you see . . . a terrestrial isopod crawl across the sidewalk, what organism is at the controls . . . the isopod, or a parasite?"

READINGS

Moore, Janice. "The Behavior of Parasitized Animals: When an Ant Is Not an Ant." *BioScience* 45, no. 2 (1995).

———. "Parasites That Change the Behavior of Their Host." *Scientific American* 250 (1984).

Raham, Gary. "Pill Bug Biology: A Spider's Spinach but a Biologist's Delight." *American Biology Teacher* 48, no. 1 (1986).

Sutton, Stephen L. *Woodlice.* London: Ginn & Co., 1972.

Sutton, S. L., and D. M. Holdich, eds. *Biology of Terrestrial Isopods.* Proceedings of a symposium held at the Zoological Society of London on July 7 and 8, 1983. London: Oxford University Press, 1984.

Tack, Stephen L., and Arlan L. Edgar. "Reactions of Michigan Isopods to Various Conditions of Moisture and Light." *Michigan Academy of Science, Arts & Letters, Papers* 51 (1966).

4

I MENTIONED some pages back that my brother, Bil Gilbert, is the writer in the family who takes an interest in vertebrates. He kindly leaves the invertebrates, all millions and millions of them, to me. I've always thought I had the better end of the deal, and I believe he thinks he does, so we have a fond, happy, and informative way of getting along.

One summer day when I was visiting Bil and his wife, Ann, at their home in southern Pennsylvania, we were sitting out on their patio looking at the stream that meanders through their front yard and talking over matters zoological. Bil told me about the crow he had been raising from a nestling, who would soon be old enough to hack back into proper wild life, when the conversation shifted to millipedes, those hard "worms" with legs that Bil and I used to call thousand-leggers when we were children growing up in Michigan.

More summers than not, Ann and Bil's house is invaded by millipedes, which make quite a nuisance of themselves. It isn't that individual millipedes do anything bad. They don't bite or eat clothes or nibble on sills and beams. About the worst thing they do is give off an unpleasant odor. But thousands upon thousands of them creep under the kitchen door, find their ways through cracks and crevices, and crawl up the walls, and each morning Ann has to sweep up the night travelers by the bushel basketful. All the neighbors along the creek have similar invasions during millipede summers.

Millipedes come in a variety of sizes. Ann and Bil's were a couple of inches in length, hard to the touch, and they coiled up when disturbed. Their tiny paired legs ran the length of their body, looking like a row of bristles. They were invertebrates, to be sure, so Bil started peppering me with questions. He knew they weren't worms or insects, but what exactly were they? Were there different species? What were they up to? What did they eat? Why did they appear in such hordes from time to time? Lemmings, he knew, massed in periodic irruptions, a behavior apparently caused by some sort of social irritability factor. Could the same be true of millipedes?

I couldn't answer his questions. All I could tell him was that millipedes belong to a class of invertebrates all their own, as do their distant cousins the centipedes, the sinuous orange creatures with fewer but longer legs, which you mustn't pick up when you find them under a rock because they can deliver a nasty dose of venom. Millipedes aren't insects, class Insecta, or spiders, class Arachnida, because they have too many legs; no, they are in another class, but at the time I couldn't remember the name.

A few weeks later, at my request, Ann gathered up a couple of

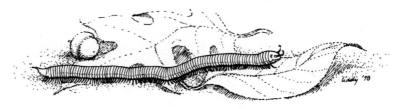

A millipede (1⅔ × lifesize).

stray millipedes from the kitchen floor and popped them into a pimiento jar filled with 70 percent alcohol solution to kill and preserve them. I'd begun reading up on millipedes and had made an appointment to take her samples to the Smithsonian Institution's Natural History Museum, where over the years I've been helped by a number of kind invertebrate zoologists to identify and learn about various puzzling creatures. Puzzling creatures are many and human creatures are few, particularly invertebrate-specialist human creatures. The Smithsonian has one of the best invertebrate collections in the world and a research staff without equal, but it does not have a specialist for each and every invertebrate. However, I'd found that the researchers could usually put me in touch with someone knowledgeable if no one on the staff was working with the particular animal I was interested in. That turned out to be the case with the Gilberts' millipedes.

I was met at the museum by Ron Faycik, of the Invertebrate Section. He looked at the millipedes in the pimiento jar, shook his head, and grinned. "With invertebrates around here," he said, "we more or less divide them up into those with six legs and those with more. These have a whole lot more, so I'll call up the spider guy and have him come take a look." After making the call, he said, "While we are waiting, let's see what we can find." He started pulling out metal drawers containing jars of millipedes preserved in alcohol, described, and labeled, which researchers use for study. I'd read that the Smithsonian had one of the world's outstanding collections of millipedes, but still I was startled to see how many sizes, shapes, and colors of them floated there in glass jars. Some were even smaller than mine, while others, big, fat, colorful ones, were coiled into huge jars. Ron and I had found nothing that looked like my samples when the spider expert, Scott Larebar, found us.

"I don't know *anything* about millipedes," he said, "except to tell you who does. Rowland Shelley from the North Carolina museum comes up to work on our collection now and then and do some identifications. He can probably help you out." He gave me a telephone number.

I called Shelley, who agreed to identify my specimens, so I sent them off to him in their pimiento jar. I then settled down to read the scientific papers about millipedes and a 1991 book, *The Biology of Millipedes,* by Stephen P. Hopkin and Helen J. Read.

Millipedes are part of a grouping called Myriapoda, which includes not only the millipedes, class Diplopoda (the double-legged), but centipedes, class Chilipoda (the foot-jawed, from the Greek *cheilos,* lip, because the foremost pair of legs is jawlike), and a couple of other classes of many-legged small animals. People who study them are called myriapodologists, and they meet regularly at the International Congress of Myriapodology in Paris. They like to remind people that there are a lot more things creeping around in the world than just insects and spiders, which they believe get all the attention. Hopkin and Read write, defensively, "Millipedes are fascinating, and dare we say it, endearing creatures."

Millipedes do not literally have one thousand legs. The record-holding species has 750; most have fewer than 100. Millipedes are of the class Diplopoda because all of them — no matter their order, family, genus, or species — have bodies with doubled but fused segments, from each of which grows a pair of legs. When they hatch, infant millipedes have only a fraction of the segments they will have as adults. They add segments and, consequently, legs, as they grow with each molt. A typical millipede has a hard, tough, mineralized cuticle, ideal for an animal that burrows through the upper layer of soil or leaf litter and feeds on decomposing leaf scraps. By the time it becomes an adult, it has consumed five times or more its own weight, in the process helping to turn vegetable debris into soil and generally tidying up the world.

As invertebrates go, millipedes are fairly long-lived. Although some grow, mate, and die within a single year, individuals of many species live for three or four years. Some, given a sparse food supply, can live even longer — eleven years is the record in captivity. (Recent studies have shown that many slightly underfed animals live longer and postpone developmental stages; there's not much point in reproducing when resources are scarce.)

The sexes are separate in millipedes, and species vary considerably as to how they comport themselves to assure new generations. Their sexual parts, typically, are a few segments behind the head and on the underside. Female millipedes have a specialized reproductive opening to receive the male's sperm, which is stored in spermathecae until it is time to fertilize the eggs. In some species females take up sperm that males secrete in droplets upon a web spun on the ground. In other species the male takes a more active role, clambering up his partner's body and wrapping himself in a coil around her until their reproductive segments meet. Joined to her in this fashion, he uses a pair of modified legs called gonopods to transfer his sperm to her. Mating may last from a few minutes to several hours. After they have separated, she fertilizes the eggs and lays them. The eggs are protected in a variety of ways, depending on the species, ranging from a simple coating secreted from the mother's body to an elaborate nest. Young millipedes hatch out of the eggs and begin to molt and grow and add segments immediately, relying on stored energy from the yolky maternal egg to take them through several molts, until they are able to feed and live independently.

The males of some species exhibit a curious bit of biology called periodomorphosis, a midlife crisis of sorts in which a sexually mature male reverts, by molting, into an immature stage with undeveloped sexual parts but retains the ability to molt and mature again. The function of this step backward is a puzzle to myriapodologists, who suggest a number of theories: it may be an adaptation to extend reproductive life in times of stress or scarcity, a way to conserve metabolic resources for additional sperm production, or a response to gender imbalance in the population. But some entirely different explanation may turn up.

Millipedes are slow-moving animals, despite their many legs, and might seem at first to be vulnerable to predators. But in fact they do not become dinner very often. Their tough, mineralized exoskeleton is remarkably strong. One experimenter has shown that an individual millipede can support 25,000 times its own weight without being crushed. And they have some remarkable defenses. Some defecate when touched. Some are bristly. Some produce chemicals that are not only antifungal, antibacterial, and stain blue but are "repugnatory," in biological jargon, defending the animal from being eaten. Of these, some are bright-colored, or even luminescent, to advertise their inedibility and to scare off predators. In addition, when disturbed, touched, or endangered, most millipedes coil, tucking in their vulnerable head, antennae, and feet. This serves not only to protect them within their hard, armored shell, but also to position in the best way possible the openings to their repugnatory glands, which secrete those smelly, itchy, blue-staining chemicals. The secretions of these mild-mannered, plant-eating animals vary according to species; they are not poisonous, and they range from offensive in odor to pleasant (at least to humans). Rowland Shelley's particular specialty, he told me, was a family of millipedes that produce a sweet odor (the family belongs to the order Polydesmida). Millipede secretions are chemically complex but include some familiar compounds: cyanide, with its almondlike odor; a terpenoid that smells of camphor; and benzaldehyde, familiar to any beekeeper who has used it to drive bees from honey at harvest time.

Whatever antibacterial, antifungal qualities millipedes have within their bodies, and whatever defensive effect their secretions and tough cuticle have, a few predators have managed to get around them. A number of diseases attack millipedes, including a

bacterial one introduced by nematodes, which suggests the possibility of a biological control for millipedes when they become
pesty. Occasionally birds, turtles, armadillos, a variety of amphibi-
ans, and some insects and spiders manage to dine upon them.
Easily the most entertaining example comes from Thomas Eisner,
of Cornell, who discovered that one kind of African mongoose
had learned how to unwrap the protective coil of a large millipede. The banded mongoose, *Mungos mungo,* Eisner writes, picks
up millipedes of the genus *Sphaerotherium* in its front paws and,
standing, hurls them backward between its legs at rocks, on which
the hard cuticle shatters and uncurls, allowing the mongoose to eat
the soft and tasty body parts within.

We tend to be impressed by numbers. So the existence of a mere
ten thousand known species of millipedes may not sound like
much when compared to class Insecta's more than a million species. From the numbers, millipedes may not seem to be one of life's
notable successes. Yet if we consider all the failed experiments that
millipedes may have witnessed during the more than 400 million
years they have been creeping around on this planet, maybe we
will think otherwise. The only vertebrates back in those Silurian
times were the fishes swimming in the ancient seas. Myriapods
probably got their start in water too. The oldest known myriapod
fossil is a centipede-like animal found in Upper Silurian sediments,
and the earliest known definite millipede is *Kampecaris tuberculata,*
in Silurian Old Red Sandstone. Somewhere back in those times,
even before there is a record of insects, millipedes emerged from
the ocean and began adapting to life on land: stepping along on
those paired legs; developing defenses against whatever strategies
other bits of life came up with in attempting to turn them into
dinner; outlasting many flashier experiments; enduring, with only
modest changes down to the present, cataclysms, mass extinctions,
climate changes.

Not long after I'd sent off my millipedes, Rowland Shelley telephoned me about them. Although to Ann and me the two had looked pretty much alike, he told me they were females of different species. Identification is usually made, he said, by noting differences in male sexual parts, so one of them, belonging to the genus *Abacion,* could be either of two species, both of which had the roughened, crested cuticle of the sample. But the other, the smoother one, was so familiar to him that he could identify even a female. Its name was *Ptyolis impressus* and it was a member of a widespread family, the Parajulidae, order Julida. It was less frequently found than the other, he said, but not really rare, something of a wanderer. It is unusual in its feeding habits; unlike other millipedes, it is omnivorous, occasionally adding animal scraps to its vegetable diet. From his knowledge of their habits, he didn't think that either of these millipedes would gather in the great hordes of Ann and Bil's millipede summers.

I had asked Bil to describe the odor of the mass invaders. His answer was "something like elderberry leaves, musty, ancient, dank." When I reported that to Shelley, he said, "Well, then, they are probably *Oxidus gracilis,* family Paradoxosomatidae, of the order Polydesmida. They are often the ones that irrupt and invade houses, and they smell like that. They've been little studied because they probably came here from Southeast Asia." He explained that there were still so many interesting things to be learned about our native species (some one thousand in forty families) that researchers would rather spend time working on them than on a species that may have made its way here in the soil of imported plants.

Rowland Shelley sighed. "People always wonder why this or some other question relating to millipedes is so little studied. There are ten orders of millipedes found worldwide, and they are very diverse in their biology and behavior; they are interesting to a myriapodologist, but maybe they don't seem dramatic to others. After all, millipedes are not harmful and have little economic importance that we know. If they did something bad, the government would spend a lot of time and money figuring out how to kill them, and in the process we might learn something about their biology."

It was a state of affairs with which I was familiar from the lack of published information about camel crickets and other interesting but harmless creatures I've busied myself with over the years. In the case of millipedes, our ignorance may even cost us something in human health. A 1981 paper by J. Tinliang and others indicated that a substance extracted from millipedes halted the growth of tumor cells for six hours. Some millipedes also produce natural sedatives.

Shelley had generously taken the time from his museum and research duties to identify my millipedes and tell me what he knew about them. I asked him how he had come to specialize in a field with few researchers. Had he studied them in college?

"How could I?" he replied in his soft drawl. "None of the professors knew much about millipedes, and what little they taught was usually wrong. No, I came to work here at the museum about twenty years ago, and the curator of invertebrates wanted to know what group I'd like as a specialty. I gave it some thought and remembered the first time I'd ever seen a millipede, a thing creeping along that looked like a cigar with legs. I asked my father, 'Daddy, what is it?' and he said, 'Son, that's a millipede.' Well, I remembered wondering about it, so I told the curator I guessed I'd like to specialize in millipedes. There were only two researchers, both of them in Virginia, and they were kind enough to let me come and work with them every year for a couple of months. I would just sit with them and ask questions until, eventually, I learned enough to go on with the work myself."

There have been many printed reports about periodic irruptions of millipedes, several of them referring to *O. gracilis,* some to other species. Some of the irruptions seem to be annual, others cyclical, still others singular. All of the reports are anecdotal, and no real science has come out of them. But in reading the reports, I noticed similarities to Ann and Bil's invasions. In nearly every case the millipedes came in the summertime in places near streams; they usually traveled at night, attracted to artificial light. One report stated that millipedes were present in such numbers that it was impossible to walk without stepping on them; another spoke of wells being filled and contaminated by millipedes as they drowned. In yet another case, farm workers were overcome with nausea and dizziness when the fields they were working were invaded. Some said the millipedes moved purposefully in the same direction; others said they milled about, occasionally wadding up together in big balls. There were stories of millipedes massing on railroad tracks ranging from eastern Europe to the American West. In some cases, sand had to be strewed on the tracks lest trains be derailed by their numbers. Possible explanations of the irruptions include population migrations of an animal that moves slowly and so takes a long time getting from here to there, dispersal to adjust the gender imbalance of a population, dispersal to favorably damp surroundings. One writer suggested that each millipede invasion had a different cause, that there was no single explanation.

When I talked to Rowland Shelley, I asked him what the current thinking was about irruptions. Could they possibly be the result of "social irritability," as Bil had suggested? He didn't think so. "Social irritability applies to animals like lemmings," he said, "and they have a lot more mobility than millipedes do and are behaviorally more complex. Social irritability would imply a more organized consciousness than a primitive invertebrate probably has, but some invertebrate equivalent of it might be described as a 'density-dependent population factor' in which crowding, depletion of resources, and a lack of mates might all be contributing factors."

In 1742 Charles Owen published *An Essay Toward a Natural*

History of Serpents, a compendium of such information as that doctor of divinity was able to find about such "serpents" as snakes, griffins, sea monsters, tarantulas, honeybees, and millipedes, which he called *Scolopendra.* That Greek word originally meant "millipede," but it has been transformed through taxonomy to become a contemporary genus of tropical centipedes. But never mind. Owen's description lumps together the two kinds of myriapod anyway and gives a hodgepodge of misinformation about a generic animal. His illustration, a drawing of which can be seen opposite, puts together attributes of millipede and centipede and throws in a little ocean worm for good measure. It is an example of pack-rat scholarship, a drawing copied and recopied by literary gents who had never seen any of those animals. It resembles a millipede in the same way that the final tale resembles the original story after it has been passed around a circle of ten-year-olds playing Rumor. The many paired legs have been transformed into feathers on a beastie squeezed, in Owen's original, between the Sea Serpent and the Mistress of Serpents. Here is his accompanying entry:

> The Scolopendra is a little venomous worm, and amphibious. When it wounds any, there follows a Blueness about the affected Part, and an Itch over all the Body, like that caused by Nettles. Its Weapons of Mischief are much the same with those of the Spider, only larger; its Bite is very tormenting, and produces not only pruriginous Pain in the Flesh, but very often Distraction of the Mind. These little Creatures make but a mean Figure in the rank of Animals, yet have been terrible in their exploits, particularly in driving People out of their Country; Thus the Inhabitants of Rhytium, a City of Crete, were constrained to leave their Quarters for them.

So far, Ann and Bil have not been "constrained to leave their Quarters" because of millipedes, although Ann says she has thought about it after sweeping up dustpansful of them. But our

Charles Owen's *Scolopendra*.

understanding of their massings seems to have expanded little since 1742. Ann and Bil and their neighbors along the creek have millipede summers more often than not, and it is as interesting to speculate why they *don't* have them in some seasons as why they do in others.

When we were once again sitting outside their house on the patio, drinking coffee, I relayed what I'd learned about millipedes. They had tried to correlate the lack of irruptions in the preceding two summers and occasional other ones with changing weather patterns, cycles of winter cold, and rain, but could find no correspondence. We sat quietly musing, and then I looked at Ann. She was wearing that private look I have come to know over the years. She got up and went back into the kitchen. Returning with a smile on her face, she held out on the palm of her hand a raffia figure clothed in a scrap of red cloth. "I was in Mexico a couple of winters ago, you remember," she said, "and a woman gave this to me. It is a spirit doll that she said could grant one wish. I'd been getting pretty tired of sweeping up millipedes every summer morning over the years, so when I brought it home I put it in the kitchen window and wished we wouldn't have any more millipede summers. Maybe it worked!"

READINGS

Barber, H. S. "Migrating Armies of Myriapods." *Entomological Society of Washington Proceedings* 17 (1915).

Bollman, Charles H. *The Myriapoda of North America*. U.S. National Museum Bulletin no. 46. Washington, D.C.: Smithsonian Institution, 1893.

Cloudsley-Thompson, J. L. "The Significance of Migration in Myriapods." *Annals and Magazine of Natural History* ser. 12, no. 24 (1949).

Eisner, T., et al. "Defensive Secretions of Millipedes." In *Handbuch der experimentellen Pharmakologie*, vol. 48, chap. 3. Berlin: Springer-Verlag, 1978.

————. "Mongoose and Millipedes." *Science* 160 (1968).

————. "Mongoose Throwing and Smashing Millipedes." *Science* 155 (1967).

Hannibal, Joseph, and C. Talerico. "Millipede Hording." *Field Museum of Natural History Bulletin* 57, no. 8 (1986).

Hopkin, Stephen P., and H. J. Read. *Biology of Millipedes*. New York: Oxford University Press, 1992.

Morse, Max. "Unusual Abundance of a Myriapod." *Science* 18 (1903).

O'Neill, R. V., and D. E. Reichle. "Urban Infestation by the Millipede, *Oxidus gracilis*." *Tennessee Academy of Science Journal* 45, no. 4 (1970).

Owen, Charles. *An Essay Toward a Natural History of Serpents*. London: John Gray, 1742.

Roncadori, R. W., et al. "Antifungal Activity of Defensive Secretions of Certain Millipedes." *Mycologia* 77, no. 2 (1985).

Tinliang, J., et al. "Observation of the Effect of *Spirobolus bungii* Extract on Cancer Cells." *Journal of Traditional Chinese Medicine* 1 (1981).

Wood, Horatio C. *Myriapoda of North America*. Philadelphia: Sherman, 1865.

5

WHEN BIOLOGISTS TALK about successful animals, they sometimes mean animals, or groups of them, that are very many. It is in this sense that beetles are successful, with 370,000 known species and millions more awaiting discovery. Beetles are the most numerous order of animal life. And in the class to which beetles belong, the insects, there are more individuals and more species than there are of any other animal. Tom Eisner, the Cornell scientist who discovered the millipede–hurling mongoose, has observed, "Bugs aren't going to inherit the earth; they own it now." They are small, short–lived creatures that thrive without moving far from home and produce many offspring. Those many, speedy, circumscribed lives allow the adaptive, evolutionary process — and hence speciation — to make itself apparent quickly, more quickly than among the kinds of bigger animals, which are fewer in number, take years to mature to reproductive age, and range widely.

Sometimes, however, when biologists talk about success they mean enduringness. It is in that sense that Ann and Bil's millipedes, whose origins go back to Silurian times, at least 425 million years ago, are successful. Paleozoologists estimate that more than 95 percent of all species that have ever appeared on this planet are now extinct. Local changes were too much for some. Many species disappeared during one or another of the more than half-dozen global catastrophes that ended the trilobites, the dinosaurs, and a whole bunch of shelly animals. Zoologists also tell us that the animals that manage to endure a severe period in which many contemporary species are extinguished are often tough — or lucky — enough to survive subsequent mass extinctions.

Little boys and girls who love dinosaurs will promptly tell us that the dinosaurs were extinguished some 65 million years ago at the end of the Cretaceous, when something awful happened to the planet; perhaps an asteroid struck it and threw up such a cloud of debris that the earth became dark and cold under it. The millipedes survived that.

Long before dinosaurs even existed, 250 million years ago, a much greater extinction of life occurred. No one knows what caused that event of the Permian period, which closed the Paleozoic era, the era of life's dawning. However, there are hints that the causes may have been chemical. Perhaps enormous numbers of erupting volcanoes spewed out poisonous gases that altered the composition of sea and air. A recent theory suggests that a massive release of carbon dioxide from the ocean depths killed many marine animals outright and produced what we now call a greenhouse effect, making the land too hot for many animals. The millipedes survived even that. They survived several other mass extinctions, too.

That kind of success doesn't mean that millipedes of the species Ann gave me in the pimiento jar, *Abacion* sp. and *Ptyolis impressus,* were crawling about in Silurian soils along with the oldest known millipede, *Kampecaris tuberculata.* Particular species come and go, have their own life spans. Those spans are represented by very guessy figures, but some respectable authorities estimate that the average life span of an animal species ranges between 5 and 10 million years. Many marine invertebrates fall within that range, as far as can be documented, but we humans don't. Several human species have become extinct in the few million years something recognizably human has been around. Our own species, *Homo sapiens,* hasn't been here very long, to be sure, and no one knows how long we will last, but we are mammals and, so far, mammalian species have tended to die out at about a million years of age on average. It is easy enough to find fault with those numbers, which are silly in many ways, but the larger point is that species do come to an end. However, species belong to big familial clumps of other animals to which they are related in evolutionary descent, biochemistry, bodily form, or life plan in varying degrees of closeness. By convention we call these groupings by the collective terms genus, family, order, class, and phylum. So even though the pimiento-jar millipedes have not been around for 425 million years, members of their class — things millipedish — have.

There are even more successful animals than millipedes in the sense of enduring. They are various marine invertebrates, which had, of course, a head start on the ones that crawled out onto the land, since life first began in the sea.

One group of these marine invertebrates is the sponges. They survived not only all the mass extinction events that millipedes did but also the one that ended the Cambrian period, a cataclysm so severe that only a few groups of animals survived. There were sponges around building reefs before corals ever invented themselves. There were sponges around before other marine animals developed shells and prickles to defend themselves. There were sponges around when potential Maine and potential Britain were still tucked together at more southerly latitudes. Sponge kinds of animals date back more than 600 million years. We know this because there is a record. Sponges have a skeleton, so they fossilize. You can have a skeleton if you are an invertebrate, you just can't have a backbone. Sponge skeletons are made of slivers of mineral called spicules and/or tough, fibrous strands of protein called

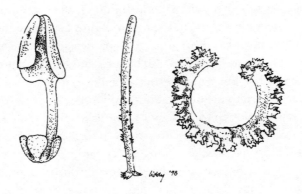

Typical sponge spicules, drawn from scanning electron
microscope photographs (enlarged many times).

spongin. The spicules vary in shape and structure according to
species and in fact are used to identify them, since the otherwise
rather blobbish body of a sponge can vary in color and shape depending upon its living conditions. Scanning electron microscope
photos of the spicules show them to be architectural, fanciful,
beautiful.

One of the animals living between the tides at the park near my
home in Maine is a sponge. Its common name is descriptive:
crumb of bread sponge. It looks like a bread crumb, perhaps a
moldy one, in shades of green, tan, and yellow. Its scientific name
is *Halichondria panicea,* which means the same thing. Bread-crumb
sponges are flat in form and spread irregularly around the rocks.
Their most prominent features are bumps with holes in the tops,
like tiny volcanoes.

I wondered how these animals get on in the world, so one time
while I was in Washington, I went to the Library of Congress to
find out. I located a paper by Paul Fell and William Jacob, who
had studied the reproductive behavior of various species of the genus at the Mystic River estuary in Connecticut. The population of

Halichondria there varied in gender, blurrily male, female, and hermaphrodite, a condition known as "incomplete gonochorism." And, clubby animals that they are, they aggregated so tightly as larvae that they sometimes blended and became genetic mosaics. They must have been exasperating to study. The authors wrote that the various species were not clearly defined, that taxonomic work needed to be done on the group before anyone could tell who was what with any exactitude.

Their paper raised a lot of questions in my mind, so I set out to find what I could about sponges in general. Sponges are pretty basic, about as simply organized as any many-celled animal we know. They are not quite as basic as those blue, pink, yellow, and green cellophane-wrapped rectangular things on the grocery store shelf, which of course are plastic. People still do buy, on occasion, irregularly shaped tan natural sponges for bathing or washing cars. That is about as close as most of us ever get to a real sponge. But these are mere skeletons of a certain few animals that, when alive, looked something like a hunk of raw liver. Sponges are biologically interesting, because they represent not only a transition be-

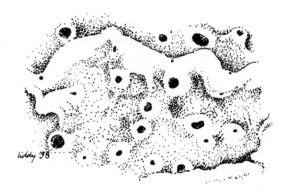

A community of crumb of bread sponges,
Halichondria panicea (lifesize).

tween unicellular and multicellular life, but a form that is unique. The organization of their bodies, simple as it seems in comparison to a millipede or a periwinkle, is unlike that of any other animal. In an evolutionary sense their kind of organization never led to anything else. Sponges are a surprise, an evolutionary sideshow, and some taxonomists have thought them so special that they have put them in a separate subkingdom of animal life.

When they are alive, sponges are beautiful, come in brighter and more varied colors than those cello-pack rectangles. They are blood red, purple, green, brown, snappy yellow, blue black. But they are calm, nerveless, living out their adult lives attached to hard surfaces under water. As a result, in the past they were believed so lacking in animal pizzazz that they were classified as plants. Because of the difficulty of entering their world — many sponges live in deep, cold water — few people had the chance to see them alive. Those taken out of the water quickly died, their colors faded, and the complex way they lived remained little known.

You and I know perfectly well what we mean by the words "animal" and "plant," but reality has never felt obliged to conform to human speech. Some animals, sponges, for instance, but also others such as sea anemones and the hydrozoans upon which nudibranchs feed, are hard to fit into our commonsense definition of animal because they seem as fixed to one spot as any rooted plant. Biologists make distinctions having to do with embryonic development, metabolic activity, and physiology. But a respectable rule of thumb, not far from common understanding and also not far from science, defines animals as organisms that consume food made by others (biologists call them heterotrophs) and plants as those that make their own food from materials at hand (autotrophs), ordinarily using the sun's energy, through photosynthesis, as the engine of transformation. Even such carnivorous plants as Venus flytraps are basically photosynthetic factories that produce food. The meaty tidbits they capture are just the cherries on top of the sundae, and they can survive quite nicely without them. It is in this sense, the manner in which they get their food, that sponges are animals.

It has only been one hundred years since the skeptics conceded that although sponges didn't chase about and capture prey, they did gobble up food, not manufacture it. The proof is that colored water placed near a sponge gets swept into it, cleaned of any suspended food bits, and pumped back out. The bumpy openings in the bread-crumb sponge that reminded me of little volcanoes are the vents from which the colony of sponge animals flushes waste water.

Sponges are often described by what they lack. Not only do they lack liver and lights, they have no real organs at all. They have no eyes, no ears. They don't even have a nervous system, and any coordination or even reactivity that they have depends on messenger substances diffused throughout their bodies and on individual cell jostling against individual cell. They even lack stomachs. Instead, they are layered masses of specialized cells that function in a seemingly muddled way. Yet their way of life is efficient and complex. And they have been at it for a much longer time than most other animals that we think of as having exciting lives.

The outside layer of a sponge's body is pierced with many pores (the phylum name, as a result, is Porifera). Those pores are mouths that lead into the central mass through runnels lined with specialized cells with sticky collars and hairlike flagella, which wave and beat through the sponge body the nutrient-laden water that has entered the mouths. Sponges are filter feeders. Those nutrients, many of which are too small to be seen, are trapped by collar cells, and special amoeba-like roving cells carry them to other parts of the sponge body. The water, stripped of its nutrients and beginning to pick up the chemical wastes of metabolism, is swept into internal cavities and out the large vent to the water outside. A

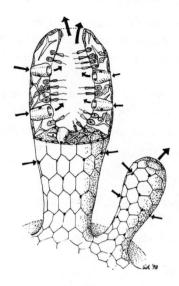

Diagrammatic representation of a pair of sponges showing
their cellular structure, including the sticky collar cells with
their flagella, and the flow of water in, through, and out.

sponge is an animal with mouths, mouths, mouths; one open end;
and no internal fluid except for all the sea itself. So strong is the
current rowed through a sponge by those waving, albeit uncoor-
dinated, flagella, that in the Caribbean a colony of sponges filters
the entire column of water above it through their bodies each day,
scrubbing it clean, and in its circulation altering the watery environment.

Sponges can reproduce asexually by a process something like
budding. They can also reproduce sexually. Sponge sex seems so
haphazard that it is hard to believe that it works, but there are
sponges everywhere — an estimated ten thousand species, only
half of which are named and known — in cold waters and warm,
in fresh water and salty seas, so it apparently does work. Some
sponges come in separate genders; some are hermaphrodites. Dur-

ing mating season, male cells produce sperm, which float out of the sponge's vent in great clouds, making the water above it murky. As the sperm waft through the water, some are eventually picked up through another sponge's mouths and fertilize the eggs produced within.

The free-swimming larvae that grow from this mating have been little studied, but in at least one species the larvae have been found to produce serotonin, which is not present in adults. Serotonin plays a role in cell biochemistry, but we think of it primarily as a neurotransmitter. What is a neurotransmitter doing in an animal that lacks a nervous system? And why isn't there any serotonin in adult sponges? The researcher who discovered this interesting aspect of sponge life no longer has funding to continue his work, but someday another researcher may pick up the puzzle.

For all their cell specialization, sponges are still just a collection of cells, not organs. A living sponge pressed through fine silk separates into individual cells but can reorganize by progressive association and become a new sponge or sponges, none the worse for wear.

Within the simple basic sponge structure there is considerable variation of form, folding, and spreading of cell masses. Some sponges are shaped liked boulders, some like vases or candelabra or baskets. Some, like the crumb of bread sponge, spread across surfaces in colonial groupings. And what goes on within those layers of cells is quite complicated.

The Smithsonian has one of the few spongologists in the world, Klaus Ruetzler. He and I had lunch together in the staff dining room of the Smithsonian's Castle one rainy November day. Klaus is a bespectacled man with gingery beard and hair. He speaks with an Austrian accent and has a kind, but formal, European manner. In addition to being the resident sponge specialist, he is director of the Smithsonian's Caribbean Coral Reef Ecosystems Program, which maintains a research station at Carrie Bow Cay, an island of less than an acre on the 200-mile-long barrier reef off the coast of Belize. The reef, a United Nations World Heritage site, is surpassed in length only by the Great Barrier Reef of Australia and a few others in the Pacific. As sponge talk extended our lunch from one hour to two and then three, and as cold rain lashed the roof, I longed to see the spot where Klaus conducted his research. He generously said that if I could get to Belize in the late winter he would show me around his spongy domain.

In Belize a few months later, I checked into lodgings at a cay near Carrie Bow, and one sunny afternoon I hitched a ride in a passing divers' boat to Klaus's domain, stepping out on the dock on a white sand island with palm trees.

It was twenty-five years earlier when Klaus and another Austrian marine zoologist, Arnfried Antonius, discovered this beautiful island. They happened upon it with a boatman who had lost his way and was trying to find a passage through the reef. Intrigued by the tiny island and the buildings on it, they drew up to the dock. Only pelicans were there, but a sign read (as it still does) "Welcome to Carrie Bow Cay." One of Klaus's missions on that trip had been to find a place that would serve as a base for the Smithsonian's marine research, and Carrie Bow Cay seemed ideal. As it turned out, the island was owned by a family of dedicated environmentalists. They worked out a lease, and the Smithsonian station was established. To mark the station's twenty-fifth anniversary, Klaus was joined in February, when I visited, by his old friend Arnfried, from the University of Vienna.

Arnfried, who had not been back for seventeen years, was the discoverer, right at Carrie Bow Cay, of black band disease of corals. Coral reefs are home to animals from nearly every phylum, the base of an ecosystem so important that reefs are called the tropical forests of the seas. They are now threatened and dying around the world, probably because of the degradation and warming of the waters. So any disease that affects them, such as black band, is of

Carrie Bow Cay.

enormous environmental consequence. On this visit Arnfried was taking measurements to discover how much this area, pristine when the two men found it twenty-five years ago, has been altered by human use. Those years have brought exploration for oil offshore, increases in tourism and fishing with power boats, increased use of agricultural chemicals, more industrial pollution, and silt flowing down rivers and into the ocean.

The station is able to house comfortably six researchers at a time; most come for a few weeks, gather data, then return home. Over the years hundreds of researchers from the United States and abroad have come here and have gradually built up a sizable published record of an intensely studied area, turning it into "one of the classic locations in marine biology," according to Jürg Ott, editor of the journal *Marine Ecology*. During the week of my visit, its residents were (in addition to the pelicans) Klaus, Arnfried, and Ilka Candy Feller. Candy, a researcher from the Smithsonian's Environmental Research Center in Edgewater, Maryland, specializes in mangrove studies at nearby Twin Cays, where Klaus has also investigated the sponges that encrust mangrove roots. As the tides ebb, sponges protect themselves against the drying air with complex physiological defenses. To a sponge, water is all, but some kinds are better than others at tolerating its loss, and Klaus studies that tolerance.

Because of Klaus's stubborn protection, the buildings on Carrie Bow Cay, modified to provide laboratory space, have retained much of the original character of the farmhouse that Henry Bowman moved to the island in the 1940s to make a summer house for his wife, Carrie. Cheerful yellow siding covers the house and outbuildings. The floors are still paved with pretty tiles salvaged from a ship that wrecked upon the reef. A generator supplements solar battery power, providing the energy to run microscopes and the other equipment needed to do modern research. There is both a wet lab, with a seawater tank, and a "dry" one, dry because it is in the driest room on this tiny island, where fierce storms and hurricanes drive rain horizontally from the east.

Klaus, tanned, dressed only in bathing trunks, looked incredibly fit for his sixty years. He proposed that we snorkel out to a nearby reef so I could see not only his kingdom by the sea but the one under it, too. I had never snorkeled and was an indifferent swimmer, but I strapped on my freshly acquired mask and flippers and plunged in behind him. He turned into a merman and rapidly disappeared from my view. Eventually I caught up with him, but I was out of breath from the effort. Klaus dove down into the passages among the corals, being careful not to touch or harm them, and pointed out the sights. I watched, aghast, as bubbles of air floated up from the end of his snorkel underwater. I was trying to keep the water out of mine and it above water. I am no mermaid.

This first effort and a more leisurely underwater visit to another reef a few days later made me realize that the world underwater is the most foreign place I have ever visited. I'd seen pictures, of course, but pictures are static. Nothing had quite prepared me for the animality of this world or so challenged my notion of landscape. The rocks are animals. The trees are animals. The flowers are animals. Animals, all waving tentacles, pulsing and swaying as water moves through and around them. The meadows are algae in forms as varied as any prairie meadow. Bright fish are the spring warblers, and even more avian are the rays, flapping like birds of prey. Everything was in motion, going about the business of living, and I was a timid, craven stranger. Klaus, darting about, taking delight in all, looked suspiciously at home. He took pity on my ineptitude and offered me a supporting hand, but I was overcome, not tired. We swam back to land. After a restorative shot of rum, I asked Klaus about his ability in the water. "My mother was a daring swimmer," he told me, "and probably because she was, as a child I didn't want to swim. My parents were always trying to get me into mountain lakes, which I didn't enjoy."

Klaus was born in Innsbruck and grew up interested in animals and their ways. He planned to follow in the path of the great Austrian zoologist Konrad Lorenz and study the behavior of vertebrates. But Jacques Yves Cousteau was one of his heroes, and he also read the popular writings of William Beebe, the American zoologist who described so colorfully his descent into the ocean depths in a bathysphere. When Klaus was a student at the gymnasium, his imagination was taken by underwater exploration. "After the war, in Austria," he told me, "there was no diving gear available, so I made mine, inventing it as I went along." He would hitchhike to the Adriatic coast, try out something he'd fashioned, go back home, and rejigger it. His first underwater camera was sealed into an aluminum pressure cooker. He designed a set of goggles and fins and had them made in a home rubber press. He met other young divers, some of whom taught him to make an oxygen rebreather out of a rubber bag and a stew pot. He was carried along by a young man's love of gadgetry, but underwater he began to see things he wondered about, most notably sponges growing in caves. At the University of Vienna he asked his biology professors about what he was seeing and photographing. They had no answers, but suggested that he do his doctoral thesis on sponges. Upon its completion, he was hired by the great German-born biologist Ernst Mayr to come to Harvard's Museum of Comparative Zoology to elucidate the sponge collection there. In 1965 he went to the Smithsonian's Museum of Natural History, where he has been ever since.

Before Klaus and other young marine zoologists developed proper diving equipment and began to do field studies of living sponges, museum specimens had been mechanically plucked from their holdfasts and often damaged in the process. Pickled for preservation, they quickly lost color and degenerated. The taxonomic work done on those damaged specimens, Klaus said, was "terrible, terrible. Today Porifera remains the phylum most in need of taxonomic revision."

Of course, commercial diving for the dozen or so species of sponges used for bathing goes back at least four thousand years to Cretan civilization. Klaus relates that sponge fibers were found in the grave of Philip II of Macedon, who was stabbed to death. Philip, father of Alexander the Great, had taken a new consort and put aside his wife. The scorned woman and her son were suspected of arranging the murder to assure that Alexander would retain the crown. There is no whisper in the story of the identity of the kind person who pressed a sponge into the king's wound.

For thousands of years, when people wanted sponges for first aid and for bathing, free divers risked their lives to harvest them. But their quick descent into cold waters, limited by the length of time they could hold their breath, didn't allow them to note the niceties of sponge biology.

Sponges, builders of reefs before there were corals, are still an important part of the world's ecosystem. "Although," said Klaus, "if they hadn't survived all those mass extinctions, something else would have, and we would have a different ecosystem." Sponges are serious recyclers of the smallest food particles, as well as hospitable housing for a wide variety of other animals, including worms, brittle stars, fishes, and shrimps, to name a few. When

J. Emmet Duffy, an ecologist from the College of William and Mary, investigated one species of snapping shrimp at Carrie Bow Cay, he found colonies of as many as three hundred shrimp inhabiting a single sponge. The colonies were eusocial, meaning that they were organized into castes: one reproductive queen and an entire population of offspring that share food and defend the colony. The interesting thing about the discovery is that apart from one group of peculiar mammals, the mole rats, only insects had previously been found to be eusocial.

Few animals feed on sponges. A sea turtle may take a bite now and again, but they are full of those splintery spicules and relatively low in nutrition. Most importantly, sponges have a wide array of chemical defenses, which they advertise with what we think of as foul odors. These chemicals are interesting to biochemists, who in recent years have begun to investigate their potential medical and commercial uses. Some sponges, when touched, cause a painful rash. I spent one day with Candy Feller in the mangroves, helping her mark out some of her experimental plots. As I leaned over the boat's prow to tie a rope to a mangrove root, Candy cautioned me not to touch the bright red sponge, slightly camouflaged by algae, that was growing on the root. It was a fire sponge, one of those Klaus had studied; it would hurt me if I touched it.

Nearby, fastened to another root, was a tunicate. Although it appeared spongelike to the undiscerning eye, it was something quite different. Tunicates are of special interest to those of us with backbones because, although they are technically invertebrates, they have, at least during embryonic development, a notochord, a flexible supportive rod. Very like a backbone, but not yet bone. They belong to the phylum Chordata, as we do. Zoologists think that something like a tunicate was ancestor to all of us, vertebrates and modern tunicates together. I gave the tunicate, dim and dun, a friendly nod and saluted our common parent.

Candy and I waded through shallow open ponds in the center of the mangrove island, stepping carefully around stinging hydroids, the water sloshing around our ankles. She pointed out other sponges, almost hidden in the sediment, pulsing nutritious water in, waste water out, their animality obvious. Once she stopped and said, "Oh, look! A nudibranch. I think they are the coolest animals." She scooped up a different species of the animal I'd been looking for in Maine. This one was pale ivory, an inch and a little more in length, with lovely filamentous projections along its back. She put it back, and I watched it until it disappeared into a swirl of sediment.

We all know that animal life depends on plant life. A cow must graze on grass before we can eat it or have its milk. The squirrels my Ozark neighbors kill for stews must first feed on acorns. Plants, through photosynthesis, transform the earth's chemicals that all of us, herbivores and carnivores alike, depend on for food. On the land the plants are obvious. And northern seas are murky because of their rich layers of phytoplankton, those floating plant particles that serve as graze for fish and other animals. But tropical seas are clear, beautifully blue and azure. They are clear because they are poor in phytoplankton, yet tropical waters are full of animal life: grazers, carnivores that eat the grazers, and innumerable filter feeders. Where, then, are the plants that are the base of their economy? There are beds of turtle grass in the shallows, so grasslike that it is a surprise to learn that it is an alga. There are other algae, too, in an array of plantlike forms, making gardens for the grazers that live there. Candy kept stopping on our watery hike to point out pretty algae — mermaid's wine glass, green strands topped with dainty green vases, and merman's shaving brush, small, green, bristly. Later I found a beached, bleached bundle of *Halimeda opuntia,* one of the most common kinds of algae in those warm seas. It sits in a bowl on my desk as I write, a mass of scalloped, calcified disks. It grows in sea grass beds and, broken to bits, makes up a good part of the brilliant white sand beaches in Belizean latitudes.

Algae may be the gardens of some marine animals, but others, sea anemones, sponges, and corals among them, carry their gardens within their own bodies. Individual coral animals are polyps with waving tentacles that pull in floating food from the seas. Among the reef-building corals, an individual coral attaches itself to a substrate and to its neighbor with limy cement, and its neighbor attaches itself to its neighbor with limy cement and its neighbor attaches . . . and in a couple of thousand years you have a coral reef. In feeding, these corals pull in floating organisms called zooxanthellae, which are a part of the sparse plankton. The name, a lovely one, pronounced *zoo-zan-THELL-ee,* comes from the Greek words for animal and yellow because they stain yellow and live inside animals. They are dinoflagellates, tiny leafy algae, some far cousins of which are infamous for producing the dread red tides along coastlines, which poison shellfish.

Zooxanthellae give reef-building corals their characteristic greenish yellow-brown color. They benefit from living inside the corals, feeding off metabolic wastes and producing extra nutrition in return. They also help reduce the levels of carbonic acid, which can dissolve the coral's limy skeleton. This is a happy enough arrangement for both coral and zooxanthellae as long as the ocean is clear enough for sunlight to penetrate and fuel the zooxanthellae's photosynthetic engines and as long as temperatures are stable enough for them to be comfortable. But all over the world, oceans are becoming murkier, more contaminated, and warmer. That may be the reason, although the process is not thoroughly understood, why corals everywhere have begun to expel their zooxanthellae and lose their color — the famous coral bleaching that we read about in newspapers. The waters around Carrie Bow Cay had not been affected until the summer of 1996, when the corals bleached even there. If conditions return to what these water creatures find proper, the coral again take up floating zooxanthellae from the plankton and recover. But if the loss is extended in time, the coral sicken and die even though they can still feed themselves directly. Over the long time of their association they have become dependent on the zooxanthellae to complete their nutrition. When the coral die, the reefs fragment and crumble, and millions of other animals lose their homes.

A few sponges also harbor zooxanthellae, but more typically they grow a more ancient form of internal garden. Those living at depths penetrated by sunlight contain cyanobacteria (sometimes called blue-green algae), which produce sponge food from metabolic sponge waste in a fashion analogous to that of the zooxanthellae within the corals. And here we are, indeed, among strange bits of life, as close as possible to the beginning of it all. Cyanobacteria, lacking a proper cell nucleus, are called prokaryotes (as distinguished from those of us who are made up of nucleated cells and are therefore called eukaryotes). Even back in the time of Original Things, cyanobacteria were photosynthesizing. They may have been the first life form to start producing oxygen, which this planet lacked in significant amounts. In photosynthesis, carbon dioxide, which was abundant in early days, is transformed by the sun's energy into carbohydrate, releasing oxygen as a waste product.

Ralph Lewin, the biologist who versified upon periwinkles, is a specialist in algae. With a bow to A. I. Oparin, the Russian biochemist who thought about life's beginnings and devised the theory of the primordial soup from which it came, Lewin wrote the following in 1977.

In the Beginning

In the beginning the earth was all wet;
We hadn't got life — or ecology — yet.
There were lava and rocks — quite a lot of them both —
And oceans of nutrient Oparin broth.
But then there arose, at the edge of the sea,
Where sugars and organic acids were free,
A sort of a blob with a kind of a coat —
The earliest protero-prokaryote.
It grew and divided: it flourished and fed;

From puddle to puddle it rapidly spread
Until it depleted the ocean's store
And nary an acid was found any more.

Now, if one considered that terrible trend,
One might have predicted that that was the end —
But no! In some sunny wee lochan or slough
Appeared a new creature — we cannot say how.
By some strange transition that nobody knows,
A photosynthetical alga arose.
It grew and it flourished where nothing had been
Till much of the land was a blue shade of green
And bubbles of oxygen started to rise
Throughout the world's oceans, and filled up the skies;
While, off in the antediluvian mists,
Arose a few species with heterocysts
Which, by a procedure which no-one can tell,
Fixed gaseous nitrogen into the cell.

As the gases turned on and the gases turned off
There emerged a respiring young heterotroph
It grew in its turn, and it lived and it throve,
Creating fine structure, genetics, and love,
And, using its enzymes and oxygen-2,
Produced such fine creatures as *coli* and you.
This, then, is the story of life's evolution
From Oparin broth to the final solution.
So, prokaryologists, dinna forget:
We've come a long way since the world was all wet.

We owe a great deal — you can see from these notes —
To photosynthetical prokaryotes.

With certain species of those photosynthetical prokaryotes, the cyanobacteria, sponges have worked out a useful and amiable relationship, the kind biologists call mutualism. And sometimes, when the mass of each is just about equal, they are called not sponges but cyanobacteriosponges. Klaus has studied one of these creatures, *Terpios hoshinota,* which has gained such vigor from the partnership that it spreads rapidly and darkly across reefs in the Pacific and kills the corals.

The sponge–cyanobacteria relationship is interesting to compare to the coral–zooxanthellae partnership. One researcher, V. P. Vicente, studied sponges off the coast of Puerto Rico during the summer of 1987, when corals there were bleaching excessively. He discovered that sponges right next to the corals were little affected and did not lose the color lent by the cyanobacteria. Vicente theorized that the more ancient interaction between sponges and cyanobacteria (which he dates to at least 650 million years) may have become more stable than that of corals and their zooxanthellae (beginning a mere 200 million years ago). Sponges have been at this cooperative business long before corals ever were. Klaus agreed with Vicente's conclusion and said that his own observation confirmed that during bleaching episodes sponges are far less affected than corals.

The sponge–cyanobacteria pairing is a textbook example of mutualistic symbiosis. But nothing in the living world is ever textbook-neat. Klaus has discovered, while investigating sponges over at Twin Cays, a case of cyanobacteria and sponges not in benign equilibrium. *Geodia papyracea* is a pulpy sponge with a sandpaper-like surface that grows along the mangroves' water-covered roots where light comes through for photosynthesis. The sponge, when it is healthy, is cream or light gray or brown and lives in happy mutualism with a particular kind of cyanobacteria that Klaus calls *Aphanocapsa feldmanni*-type. When the water becomes warm, the cyanobacteria begin to multiply frenziedly. For a time the sponge encapsulates and expels the excess, but if conditions become even more favorable for the cyanobacteria, their numbers overwhelm the sponge's ability to regulate them, and the cyanobacteria become toxic to it. Klaus wrote of this discovery: "The association lacks ecological equilibrium. . . . The possibility exists . . . that we are witnessing a newly evolving mutualistic relationship in which the sponge host has not yet developed physiological mechanisms to control a balanced coexistence with its cyanobacterial symbionts."

That is something of a postcard from the beginning of biological time, reminding us newcomers, busily trying to get everything in the world under our own control and working our way, that it is not balance but change that is inherent in life, characteristic of it, even in the oldest and seemingly most stable forms.

A Sad Addendum

December 10, 1997

NATIONAL MUSEUM OF NATURAL HISTORY

SMITHSONIAN INSTITUTION

CCRE CARIBBEAN CORAL REEF ECOSYSTEMS PROGRAM

To all Participants and Sponsors of CCRE

Dear Friends:

It is a sad task to inform you that on Saturday evening, 6 December, we had a devastating fire that destroyed our field laboratory on Carrie Bow Cay, Belize. Fortunately, no-one was hurt.

Details are still sketchy but the fire seems to have started from an electrical short near the library in the main building. A strong Northerner fanned the flames which spread to the shop, wetlab, and kitchen. The fire was so hot and fast spreading that nothing could be done to stop it. . . .

Apparently nothing is left of the two larger buildings and equipment contained there, seawater system and big water tank with showers. Only the boats, compressor-generator building, small house on the south point, and outhouses were spared.

Needless to say, we must suspend most research plans until we have had a chance to regroup. . . .

Meanwhile, we shall welcome your suggestions regarding equipment items and facilities that must be replaced (specifications, priorities), including ideas for architectural design and equipping of new laboratory space and living quarters. Tell us what worked for you well in the old setup and what you would like to see improved. Let's use this small benefit of the disaster — the chance to improve — to our advantage! . . .

Thank you for your patience and support,

Klaus Ruetzler

READINGS

DeVos, Louis, Klaus Rützler, et al. *Atlas of Sponge Morphology.* Washington, D.C.: Smithsonian Institution Press, 1991.

Fell, Paul E., and William F. Jacob. "Reproduction and Development of *Halichondria sp.* in Mystic Estuary, Connecticut." *Biological Bulletin* 155 (February 1979).

May, Robert M., et al. "Assessing Extinction Rates." In *Extinction Rates,* ed. J. H. Lawton and R. M. May. New York: Oxford University Press, 1995.

Rützler, Klaus. "Low-Tide Exposure of Sponges in a Caribbean Mangrove Community." *Marine Ecology* 16, no. 2 (1995).

————. "Mangrove Sponge Disease Induced by Cyanobacterial Symbionts: Failure of a Primitive Immune System?" *Diseases of Aquatic Organisms* 5 (1988).

————. "Sponge Diving — Professional but Not for Profit." In *Methods and Techniques of Underwater Research.* Washington, D.C.: Smithsonian Institution Press, 1996.

————, ed. *New Perspectives in Sponge Biology.* Washington, D.C.: Smithsonian Institution Press, 1990.

Rützler, Klaus, and Ian G. Macintyre, eds. *The Atlantic Barrier Reef Ecosystem at Carrie Bow Cay, Belize.* Washington, D.C.: Smithsonian Institution Press, 1982.

Rützler, Klaus, and Katherine Muzik. "*Terpios hoshinota,* a New Cyanobacteriosponge threatening Pacific Coral Reefs." *Scientia Marinina* 57, no. 4 (1994).

Vicente, V. P. "Response of Sponges with Autotrophic Endosymbionts During a Coral-Bleaching Episode in Puerto Rico." *Coral Reefs* 8 (1990).

Wilkinson, Clive R. "Net Primary Productivity in Coral Reef Sponges." *Science* 219 (January 28, 1983).

6

When we behold a wide, turf-covered expanse, we should remember that its smoothness, on which so much of its beauty depends, is mainly due to all the inequalities having been slowly leveled by worms. It is a marvelous reflection that the whole of the superficial mould over any such expanse has passed, and will pass, every few years through the body of worms. The plough is one of the most ancient and most valuable of man's inventions; but long before he existed the land was in fact regularly ploughed, and still continues to be thus ploughed by earthworms. It may be doubted whether there are many other animals which have played so important a part in the history of the world, as have these lowly organized creatures.

— Charles Darwin, *The Formation of Vegetable Mould, Through the Action of Worms, with Observations on Their Habits,* 1881

BACK IN THE EARLY 1970s, when I moved to the Ozarks, it seemed as though every fifth person was an earthworm farmer. Earthworm farming was, in those days, the fashionable rural get-rich-quick scheme. People grew earthworms to sell to other people who needed breeding stock to sell to other people who needed breeding stock to sell to others. Earthworm farming was the same sort of agricultural Ponzi scheme as silkworms had been a century earlier or as Vietnamese potbellied pig-, llama-, and emu-ranching became later. But Ponzi schemes always collapse, so I wasn't sure, when recently I wanted to see an earthworm farm in operation, if I'd be able to find one. But it turns out that some people still grow earthworms for fish bait and for garden compost making.

Lori Williamson is a pretty, rosy young woman who smiles a lot. She works as a genetic counselor and lives near Mountain View, Missouri, with her husband, who teaches biology at the local high school. She is also an earthworm farmer. Her entire farm is contained within one plastic garbage pail in the laundry room behind her kitchen. Inside the pail are about a zillion earthworms contentedly turning the household garbage and leaves raked from her yard into compost simply by eating what she gives them and passing on what they cannot use. "I add a little water if they get dry," she told me, "and if I want to give them a real treat, I put in some manure." The material they cast from their bodies is dark, odor-free, rich "vegetable mould," or humus, that would delight any gardener. From that pail Lori harvests about eleven gallons of worms each year, which she doles out at a dollar per ounce to people who want to compost their own garbage. She also sells the compost for four dollars a gallon.

"I always liked earthworms," she said. "I'm interested in organic gardening and hope some day to have an organic orchard. Once I got to raising earthworms I liked them even better, so I won't sell them for fish bait, although people tell me that this kind — these are red worms — make good bait because they stay alive longer on the hook than nightcrawlers do." She wrinkled her nose in disgust at the thought of impaling one of her worms on a cruel hook. She showed me the worms, pulling back the black loam of earthworm castings, which is high in nutrients plants can use.

Lori's red worms are dainty, small, and pinkly colorful. Their formal name is *Eisenia foetida* — "foetida" because they give off a faintly garlicky odor. They are not native to this continent, but they are common throughout it. They are the sort usually farmed commercially, because in their natural state they are not burrowers but dwell conveniently in the leaf litter, so they are happy to live in Lori's garbage pail and similar setups. Nightcrawlers, the big wrigglers often used by fishermen, are also exotics, which have spread widely from the edges of streams, rivers, and ponds where they escaped from anglers. They must be harvested by hand from the wild at night. They come to the surface then because their thin skins are vulnerable to harsh, drying daylight. They are wide-ranging tunnelers and as a result do not take well to being farmed. Their scientific name is *Lumbricus terrestris*. In southern Michigan, where I grew up, we called them angleworms.

"Look," said Lori as she raked through the castings with her gloved hand, "here are some capsules." She scraped a few to one side and handed them to me. Each one, looking something like a dark grain of rice, is in effect an egg, shed by a mature earthworm. When I held one translucent capsule, or cocoon, up to the light, I could see the tiniest of wormlets beginning to grow. "Someday, if I can figure out how to sieve the capsules easily, I'm going to sell these, too," Lori told me. The red worms themselves don't over-winter, but cocoons do, and customers would be interested in buying them.

Newly emerged earthworms are made up of a snout, some body segments, and a tail, or, as we say in the trade, pygidium. I've always liked the word "pygidium." It rolls off the tongue nicely. It refers, in segmented animals, to the last bit. When earthworms hatch out, they have a sinuous body encasing a formidable, if simple, gut: a prehensile mouth, a gizzard to grind with, and a long digestive tract that absorbs all the nutritious substances that decaying vegetable matter and soil have to offer and excretes all that is not useful to a worm, lacing the whole with the rich products of earthworm metabolism.

Before I toured Lori's farm I had been reading Charles Darwin's final work, published in 1881, *The Formation of Vegetable Mould, Through the Action of Worms, with Observations on Their Habits.* Dar-

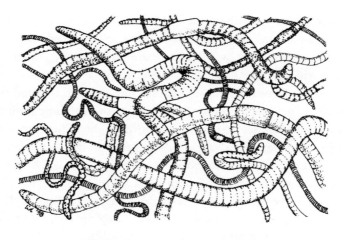

A jumble of earthworms of different species.

win was really the first observer to appreciate earthworms as makers of soil. He also grew fond of them, discovered them to be intelligent and, although blind, capable of displaying astonishing sensitivities. Up until then earthworms were often considered to be nuisances, even pests, and good gardeners eliminated them with an array of treatments and poisons. Darwin knew better. He'd watched earthworms for years, studying not only the kinds, such as Lori's, that transform surface litter into loam, but also those that burrow deep, bringing up minerals from layers of subsoil and churning it all in the process, aerating it, improving its water-holding capacity, and adding layer upon layer to the surface soil in their castings. During the time he studied them, Darwin carefully measured the amounts of soil built up slowly by earthworms and watched it gradually cover over rocks lying in pastures, then measured further as the heaped castings outside earthworm burrows eroded and rinsed down hillsides after rainstorms. Using another man's figures, he assumed that there were 53,767 earthworms per acre and that each would pass ten ounces of earth through its body each year. He concluded, from those numbers, that earthworms were important in constructing the soil on which all terrestrial life depends. Over the millennia they could even resculpt the surface of the land. Actually Darwin's numbers were on the low side. Earthworm populations per acre vary widely, depending upon the location and kind of soil, but they range from the hundreds of thousands to a couple of million. Their effect, therefore, is even greater than Darwin estimated. And all of them are passing soil through their bodies in a way that made me think of Klaus Ruetzler's sponges pumping the oceans through theirs.

Darwin, like most writers, wrote the same book over and over again. His subject was time. Time and its effects on the processes of life. Darwin brought time to biology, turning what had been largely a descriptive science into a historical one. Popularly he is thought of as the inventor of *the* Theory of Evolution. But it would be accurate to say that he published *a* theory of evolution. Evolution simply means change, and it is a fact that over life's course on the planet its forms have changed; the evidence is in the geological record. This was well known in Darwin's day and earlier; theories to explain those changes were being talked about before Darwin was even born. His contribution was a new theory about the *mechanism* of evolution. His theory of natural selection held that inborn variations that allowed individuals to compete more successfully for resources could be passed on to their offspring. Slowly, over time, the accumulation of such variations could produce new species. Darwin called his theory one of "natural" selection to distinguish it from the controlled, artificial selection that had been practiced by plant and animal breeders ever since the beginnings of agriculture.

Neither the breeders nor Darwin knew quite how these variations were inherited; Gregor Mendel did not publish the results of his experiments with pea plants until 1866, seven years after Darwin published his book on natural selection. And Mendel's work was not widely known until the turn of the century. In his experiments, Mendel demonstrated that segregated "factors" of each characteristic of a parent (today we call these factors alleles of a gene, but the word "gene" was not coined until 1909) pass into the generative units (egg and sperm) and reappear in the offspring according to certain statistical patterns. Mendel's choice of pea plants was lucky, because in a genetic sense they are in some ways simple, and the inherited factors are obvious (he'd started with honeybees, which are more complicated). As is usually the case in science, later work has shown that inheritance is more complex than Mendel's pea plants showed. Nevertheless, he laid the groundwork for what we now call genetic science.

Another theory of evolution, one that preceded Darwin's, was the one that today we call Lamarckism, a theory given wide expression by Charles Darwin's grandfather Erasmus. Jean Baptiste de Lamarck and Erasmus Darwin were men of the eighteenth century and died before Mendel's time. They assumed that characteristics acquired by an animal during its lifetime could be passed on to its offspring — that for instance, and I'm oversimplifying here, if you exercised enough, your child could be born with bigger muscles. That has proven not to be true, but contemporary thought is returning to Lamarck these days and taking another look at the inheritability of acquired characteristics. Culture, of humans and other animals, does evolve in a Lamarckian way.

Charles dismissed his grandfather's theory with the comment that it was based on "the views and erroneous grounds of opinion of Lamarck." However, Erasmus, a physician, man of science, and poet, thought about evolution long before Charles did. He might have been given more credit in the history of evolutionary thinking had he not been such an execrable poet. Charles seems to have made no comment on his grandfather's versifying. He was a kind man, and perhaps family loyalty forbade it. But young Charles must have been aware of Erasmus's florid, faun- and goddess-filled rhymes about the natural world, as well as the parodies of them. Shelley, it is true, was an admirer, but Erasmus was a radical and freethinker and Shelley admired a fellow rebel. Wordsworth found Erasmus's poetry "nauseating." Byron, in *English Bards and Scotch Reviewers,* refers to "Darwin's pompous chime / The mighty master of unmeaning rhyme / Whose gilded cymbals, more adorned than clear / The eye delighted, but fatigued the ear."

On the subject of evolution Erasmus wrote, in *The Temple of Nature,* Canto I (published in 1803, a year after his death and six years before the birth of Charles):

> ORGANIC LIFE beneath the shoreless waves
> Was born and nurs'd in Ocean's pearly caves;
> First forms minute, unseen by spheric glass,
> Move on the mud, or pierce the watery mass;
> These as successive generations bloom,
> New powers acquire, and larger limbs assume;
> Whence countless groups of vegetation spring,
> And breathing realms of fin, and feet, and wing.

He then goes on to some specifics, including:

> Imperious man, who rules the bestial crowd,
> Of language, reason, and reflection proud,
> With brow erect who scorns the earthly sod,
> And styles himself the image of his God;
> Arose from rudiments of form and sense,
> An embryon point, or microscopic ens!

So, even in rhyme, the subject of evolution and its mechanisms was in the air in Darwin's time. In 1859, fifty-six years after these verses above were printed, Charles published his *Origin of the Species by Means of Natural Selection.* What he brought to the subject was not only his careful observations made during his scientific travels, but also his interest in time and biological process. In his first scientific publication in 1842, *On the Structure and Formation of Coral Reefs,* he developed the theory that very small additions of limy secretions of individual coral animals can, over hundreds of thousands of years, build massive coral reefs. He was right, evidence has shown. And just as his final book, on earthworms, formed the basis of modern soil and earthworm science, so, too, anyone today who studies coral reefs or the ocean sciences must go back to Darwin's book as a beginning.

Darwin's genius was his ability to think of time beyond ordinary human understanding and to realize that through it small events could pile up and have big results. The idea was controversial. In 1869, after he read at a scientific meeting the short paper that years later was expanded into the earthworm book, one Mr. Fish, writing in *The Gardener's Chronicle,* raised objections. Those objections must have rankled the aging Darwin, for he recalls them in the introduction to *The Formation of Vegetable Mould:*

> In the year 1869, Mr. Fish rejected my conclusions with respect to the part which worms have played in the formation of vegetable mould, merely on account of their assumed incapacity to do so much work. He remarks that "considering their weakness and size, the work they are represented to have accomplished is stupendous." Here we have an instance of that inability to sum up the effects of a continually recurrent cause, which has often retarded the progress of science, as formerly in the case of geology, and more recently in that of the principle of evolution.

I had read Darwin's book on earthworms with pleasure. It was science writing of the old school: informative, gracefully written, free of show-off specialist language. "It was," as Sam James said, "Darwin's greatest book."

Samuel M. James is a neat, trim, tidy man with a neat, trim, tidy brown mustache and wide-awake blue eyes. He is a research asso-

ciate at Kansas State University and a professor at Maharishi International University in Fairfield, Iowa. He is also the only academic earthworm taxonomist in the United States. (There is one
other earthworm taxonomist, but he is a freelance and concen-
trates on earthworms of the Northwest.)

There are approximately 4,300 known and named species of
earthworms, and more are discovered on every collecting trip;
many of the species look pretty much alike, but they vary in biology and behavior, so anyone studying them needs a taxonomist
to tell which is which. In addition to native American earthworms, exotic species have arrived with immigrants to this continent, in the soil clinging to the roots of plants they have brought.
In recent years the spread of exotics has been rapid. Certain tough,
wily, adaptable species, which Sam referred to as "the carp of the
soil," have displaced native species. Diann Jordan, an agricultural
researcher at the University of Missouri, has been able to identify one exotic earthworm that is present in great numbers, *Apor-
rectodea trapezoides.* As to the others, native or exotic, she laughs
and says, "Only Sam James would know."

I perched on a stool in Sam's lab at Maharishi University and together, to make room for our papers, we pushed aside from the
tabletop a pile of plastic bags of pickled earthworms freshly collected from Costa Rica. Every flat surface, table, and shelf in the
room was piled with bags, jars, vials of earthworms awaiting Sam's
attention. I asked him where the best collection of earthworms
was in this country. "The Smithsonian," he told me. "They aren't
in the museum, but out at the support center in Suitland, Maryland." He went on to tell me, with a happy look on his face, of the
two weeks he'd spent there recently working on the collection.
Today's taxonomists don't simply sit in museums and identify
specimens, however. They are field biologists, and Sam is working
on a number of far-flung projects. One, with a mammalogist, is on
Philippine animals, such as the tweezer-beaked rat, that specialize

in eating earthworms. Another is a survey of earthworm populations of the Caribbean basin. Yet another concerns the expansion
of earthworm populations in North America after the glaciers of
the Ice Age retreated. Presumably there were earthworms everywhere on the continent until the glaciers crushed and froze them
under their icy weight. The glaciers, therefore, created a boundary
line for the native species, which have spread only slightly north of
it in the thousands of years since melt time. The earthworms north
of the line are exotics, which have also spread southward, invading
the territory of the natives.

I remembered how nicely Lori's exotic *E. foetida* transformed
garbage into compost (Sam observed that a native species, *Bimastos
tumidus,* could also do the job) and how Diann Jordan had found
that crops grown in land where earthworms (chiefly the exotic *A.
trapezoides*) were present produced higher yields than those grown
in land without them. "Are exotics bad?" I asked Sam. "It seems as
though they do the same stuff the natives do and, after all, since
time out of mind, animals and plants have always been spreading
into new places."

"Well, you've got to remember that native species have been
here for a very long time," Sam replied. "They've adapted to local
conditions and fluctuations in climate. They have a very close fit.
In habitat that remains undisturbed, exotics will usually not displace the native species, but where it *is* disturbed, where those
conditions to which the natives have adapted are lost, the exotics
can take over and their ecological fit may not be so good." Sam
pulled out from his files a report of research he had done in the tallgrass prairie where both natives and exotics lived. He had found
that the native species, in this case members of the genus
Diplocardia, produced through their metabolic processes a greater
amount of nutrients available to the plants there than did the exotic Lumbricidae, and that in places where the exotics had displaced *Diplocardia,* the total earthworm biomass had shrunk.

At least some of the exotic species are able to replace the natives because they can reproduce quickly and without benefit of a partner through parthenogenesis, sloughing off fertilized cocoons without mating. This is unusual among earthworms. Sexual reproduction is far more common, albeit in a hermaphroditic manner. In hermaphroditic sexual reproduction, some of the segments housing that remarkable gut specialize and become, in a single animal, egg-producing and sperm-producing areas. When earthworms pair, they come together head to pygidium. They then secrete a blanket of mucus that envelops their sexual segments, which are near the flat-looking encircling band that is the most obvious feature on a mature earthworm. It is called the clitellum, from a Latin word for saddle. They then exchange sperm, which work their way along a groove to the partner's egg-producing segment. After they have parted, the clitellum gradually begins to harden to become the cocoon, the protective capsule for the fertilized eggs, and the earthworm wriggles backward out of it, leaving the cocoon to the mercies of the leaf litter, soil, weather, and marauding small animals, mostly other soil invertebrates.

For forty-five years I've had a fondness for all those small creatures living in the upper regions of the soil. I was first introduced to them when I was a student at the University of Michigan, where I was part of a band of students who hung out with the extraordinary Marston Bates. Bates was not, physically speaking, a large man, but he filled up more space than anyone I'd ever known before. Looking back on that time, I think he made use of us groupies to talk through the books he was writing, but we didn't mind. He was a wonderful talker, probably a better talker than writer, and what he was talking about was something, shiny and new at the time, called "ecology," the study of how the parts of the world affect one another. It was biology in context, and I've always been a pushover for context. We often spent our Saturdays digging up square meters of earth in different kinds of habitat; we counted and rough-classified the animals visible to the naked eye, and we plotted population curves from our census, correlating the populations with temperature, soil moisture levels, and other factors. Once a week we'd spend an evening sprawled on his living-room floor listening to him analyze data and spin out his ideas. Ever since, I've been unable to walk anywhere without being aware of all the lives being lived out beneath my feet.

Earthworms, I learned then, were only one of a number of animals feeding under our feet. Two others have already appeared in these pages — millipedes and pill bugs — but more come to mind immediately: grubs of many sorts, springtails, mites, nematodes, and scads of bacteria and protozoans. Do they have any relationship one to another? Do they compete for resources? Do they complement one another? I read two papers written three decades apart, one in 1963, the other in 1995, that touched on these matters and reflected a change in biological outlook during that time.

Back in the days when I was digging up those square meters of soil, we subscribed to a benign view of the natural world as a place where all parts fit together, where species are engaged with one another in interlocking, necessary relationships. Peter Larsen expressed that view of the world when I asked him about the nudibranch's role in the Maine intertidal ecosystem. It is a centuries-old way of thinking. Linnaeus, who held that there were only nine thousand species of plants and animals in the entire world, observed in 1760 that he thought the planet could hardly do without a single one of them. Today biologists know that the numbers of living species, as we define them, are in the millions, although how many millions we do not know, and some scientists are beginning to suspect that "organisms have evolved through natural selection to maximize their contribution to future generations compared with other individuals of that species — not to serve functions in an ecosystem." In other words, individuals within species of the profligate natural world are many, selfish, greedy, pushy, excessive, filling up all available space, taking all the resources to their own advantage, and not all of them may be "necessary" to the functioning of an ecosystem. Some may be extras, spare parts, or, to use the currently fashionable word, redundant.

The quotation is from the 1995 paper I read, by three Swedish researchers, Olof Andrén, Jan Bengtsson, and Marianne Clarholm, entitled "Biodiversity and Species Redundancy among Litter Decomposers." The 1963 paper, entitled "The Role of Soil Animals in Breakdown of Leaf Material," was written by the grand old man of earthworm studies, C. A. Edwards and a collaborator, G. W. Heath. It reported a study of buried bags of vegetable debris, which found that earthworms were the most important members of the community of decomposer animals, but others appeared to specialize at various stages of decay, and concluded that "whether the materials were moist or dry appeared to *affect* their susceptibility by soil animals."

The 1995 study, also of buried bags of vegetable matter, concluded that decomposition rates could be predicted by variables in soil temperature and moisture *alone,* that probably some of the animals, fungi, and microorganisms were more important to decomposition than others, but that some were simply redundant, for neither their total biomass nor their diversity correlated with the rate of decay. The authors are cautious, however, about the implications of their findings, concluding:

> Key species and the diversity of functional groups may certainly be important for ecosystem processes. However, it is imperative to clearly distinguish these questions from fluffy and general statements such as "biodiversity is important for ecosystems" — a statement that has questionable support in the modern ecological literature. . . . The scientific approach is to formulate testable hypotheses concerning biodiversity and ecosystem function, and test these against data. . . . Finally, a reservation may be appropriate. It is possible that the major importance of biodiversity for ecosystem

processes is not apparent under relatively stable conditions, but that diversity is imperative for an ecosystem's response to stress or major environmental changes, such as climatic change, without any loss of ecosystem function. Perhaps rare species become important when conditions change.

Similar, if not exactly the same, studies. Similar results but different emphases. When Sam and I talked about the papers, I asked what he knew about the animals that shared the earthworms' world, how they got along.

"Very little real work has been done on the interactions of earthworms with other animals in the soil," he told me. "But those interactions are important, and we need to know about them. I can tell you one thing I have observed, and that is, if millipedes are present in the soil, there aren't many earthworms. I don't know why, but when I am digging for earthworms and come across millipedes I just start digging somewhere else because I know I won't find any there. And yes, redundancy is being talked about a lot these days, especially in considering biodiversity. I think it is a useful and helpful concept."

The great problem, of course, is knowing which species are the redundant, now or in the future, and in our great ignorance of the life histories of even those animals we have identified and named, let alone those we have not, we are a long way from being able to pin the label "spare part" on any of them. The great environmentalist Aldo Leopold once observed, more gracefully and tellingly than the Swedish researchers, that if you are taking apart a machine you do not understand, you'd better be careful to save all the parts.

I asked Sam how he came to be an earthworm taxonomist. "I graduated from Amherst in 1975," he told me, "and then I went to the University of Michigan for graduate work in ecology; afterward I visited the prairie and decided I wanted to concentrate on grassland ecology here. To understand the way prairie grasses grow, I had to know all about the soil and the role of earthworms in it. I found that very little work had been done on the topic. The taxonomic work on earthworms was old and out of date, so I had to train myself to become a taxonomist." He paused and added, "Of course, I've always liked earthworms." There it was. Sam James, Lori Williamson, Charles Darwin. People who come to know earthworms grow to like them.

A lot of people seem to like earthworms, or at least find them useful. Fisherfolk, of course, but since 1340, when earthworms first entered the Western pharmacopoeia, they have been stewed, dried, and served raw as specifics against pyorrhea, postpartum weakness, smallpox, kidney stones, and even baldness. Earthworms, being high in protein, vitamins, and niacin, recently have been investigated as a source for animal feed. However, it turns out that processing them on a commercial basis would make them slightly more expensive to feed to your dog than beef tenderloin.

People at various times and in various places have eaten earthworms and found them tasty. I have read, for instance, that the Japanese make a pie out of them. I've never written a book without a recipe in it, almost always one for pie, but I think I'll pass this time, even though I do have one for a pie that begins "Sauté two cups of whole worms." No, I'm not going to give it; nor am I going to give the recipe for Worm Drop Soup, which calls for fifteen red wrigglers; not even Spaghetti Surprise, the first ingredient of which is a cup and a half of freshly ground worms. The cookbook in which I found these and other tantalizing recipes is by B. Gayle Beadle, a name that encompasses an earthworm-digging couple. The book is entitled *The Nobody Loves Me Cookbook,* from the children's song that those of us of a certain age will remember.

Nobody loves me,
Everybody hates me,

I'm gonna eat some worms.
Long, slim, slimy ones
Big, fat, juicy ones,
Short, fat, fuzzy-wuzzy worms.

First you bite the heads off,
Then you suck the guts out,
Oh, how they wiggle and squirm,
Long, slim, slimy ones,
Big, fat, juicy ones
Short, fat, fuzzy-wuzzy worms.

Thoughtfully, B. Gayle Beadle's book provided not only those complete lyrics but also the music. The authors like worms, too, writing, "Worms make wonderful pets. It is with reluctance that the authors dress out their red wigglers. If they had eyes, it would be impossible."

Well, earthworms *are* neat little animals and, biologically speaking, of great interest. In my day the first animal that students dissected in high school biology was an earthworm. To this day I can remember laying one open with a scalpel, checking out its multiple hearts, and tracing the red, hemoglobin-marked nervous system. Earthworms represent one of the most popular styles of animal organization. They are made up of segments, as are insects. Even we vertebrates show segmentation in our backbones. Segmented animals have bodies built up by adding one like piece to another like piece, over and over again. This is such a practical way of building a body that zoologists believe it evolved several times. In worms it is especially interesting.

It turns out that an animal whose body is fluid-filled, is longer than it is wide, and has no rigid, awkward, unforgiving skeleton can be pretty good at burrowing. The fluid allows the body to change form. The most striking feature of the sea cucumbers I rescued from the pickle factory (though not worms, they are fluid-filled animals longer than wide) was their unsettling habit

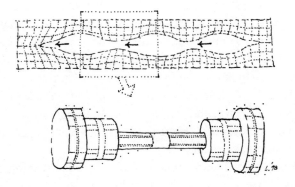

Diagram of a worm's burrowing mechanism.

of changing form while the illustrator was trying to draw their portrait.

In animals of this sort, muscles develop that can contract and extend the body, sloshing the fluid around in the process, pushing the animal through mud and soil efficiently. It is an ideal way of life for an animal that scavenges small leftovers. But if an animal that is longer than wide with a fluid-filled body is also *segmented,* as an earthworm is, with each segment sealed off from the next by a wall of tissue, the animal can be a better burrower and there-by find even more good things to eat because the muscles that control extension and contraction work with less expenditure of energy; the fluid inside each segment needs to slosh from only one end of it to the other rather than the entire length of the ani-mal. Its body becomes a series of hydraulic rams with cumula-tive effect. This, at least, is an oversimplified version of the Burrow-ing Theory of the Development of Segmentation, put forth by a British zoologist, R. B. Clark, who argued that segmentation gave the animals who developed it a competitive advantage over their nonsegmented kin.

Segmentation also allows for the physiologically efficient specialization of certain segments. Earthworms, for instance, have hearts and organs of excretion repeated in many segments, but their egg and sperm production are restricted to certain ones. And their nervous systems are regionally specialized. Darwin found earthworms to be quite intelligent, able to figure out which end of a fallen leaf to pull down into their burrows with the least trouble, a matter of wormish judgment and something suspiciously like reasoning. And they are a favorite animal of psychologists, for they can learn and make choices.

Some scientists believe that a small phylum of living animals called the Onychophora may link the worm phylum, Annelida (from the Latin word for "ringed"), to the joint-legged phylum, the Arthropoda, which includes such animals as insects, lobsters, and spiders. The linking Onychophora may show that all three phyla had the same ancestors back in the mud of Precambrian seas. Today's Onychophora live in warm, moist, tropical places and come out from under bark and stones to feed at night. They are rare. I carried a photograph of a *Peripatus,* one of about sixty-five kinds of Onychophora known to be alive today, to the rain forest in southern Belize on the trip I took to talk with Klaus Ruetzler

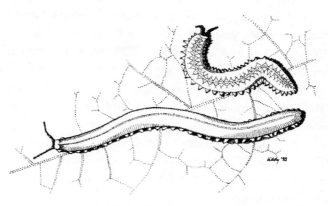

A mother peripatus and her youngster (1½ × lifesize).

at Carrie Bow Cay. I asked some Mayan Indians there if they'd ever seen animals like those in the photograph. Most said they never had, but one man said yes, and told me that their name in Mopan, the language of his people, was Ko-mes. I never saw one during the few nights I was there, although I looked for them. A student of Sam James's, however, had collected some in Costa Rica. They look like gigantic caterpillars with velvety skin. They are segmented internally, like an annelid, and their excretory system is similar to that of worms. But their circulatory system is open, much like that of insects, as is their respiratory system. They also have antennae. They walk along on plump, unjointed legs growing from each segment.

Whatever the remote creatures were that were ancestors to worms, bugs, and the shy, velvety *Peripatus,* they were probably quite different from the red worms in Lori's laundry room. Earthworms belong to the class within the phylum Annelida called Oligochaeta, which means they have just a few setae, or chaetae. Setae are the small bristles growing from each segment that give earthworms purchase as they tunnel through the soil. There are two more classes among the Annelida. One is the Hirudinoidea, the leeches, and the other is Polychaeta. If earthworms are oligochaetes because they have just a few setae, then you might expect polychaetes to have many setae, and you would be right. These mostly marine worms come with an extraordinary array of setae in often beautiful forms — fans and bristles and furs, some of fantastical texture, shape, and utility. And polychaetes are believed to be the oldest kinds of worms: some say the earthworms who traded sea for land may have been descended from them. But of polychaetes, more later.

READINGS

Andrén, Olof, et al. "Biodiversity and Species Redundancy among Litter Decomposers." In *The Significance and Regulation of Soil Biodiversity.* Proceedings of an international symposium, ed. H. P. Collins et al. Dordrecht, Netherlands: Kluwer Academic Publishers, 1995.

Beadle, B. Gayle. *The Nobody Loves Me Cookbook*. Denver: Mad River Press, 1982.

Darwin, Charles. *The Formation of Vegetable Mould Through the Action of Worms with Observations on Their Habits*, 1881. Reprint, with introduction by Stephen Jay Gould. Chicago: University of Chicago Press, 1985.

Edwards, C. A., and P. J. Bohlen. *Biology and Ecology Of Earthworms*. London: Chapman and Hall, 1996.

Edwards, C. A., and G. W. Heath. "The Role of Soil Animals in Breakdown of Leaf Material." In *Colloquium on Soil Fauna, Soil Microflora, and Their Relationships*, ed. J. Doeksen and J. van der Drift. Amsterdam: North Holland Publishers, 1963.

Ghilarov, M. S. "Darwin's *Formation of Vegetable Mould* — Its Philosophical Basis." In *Earthworm Ecology and Biogeography in North America*, ed. Paul F. Hendrix. Boca Raton, La.: Lewis Publishers, 1995.

James, Samuel M. "Soil, Nitrogen, Phosphorus, and Organic Matter Processing by Earthworms in Tallgrass Prairie." *Ecology* 72, no. 6 (1991).

———. "Systematics, Biogeography, and Ecology in Nearctic Earthworms from Eastern, Central, Southern, and Southwestern United States." In *Earthworm Ecology and Biogeography in North America*, ed. Paul F. Hendrix. Boca Raton, La: Lewis Publishers, 1995.

James, Samuel M., and T. R. Seastedt. "Nitrogen Mineralization by Native and Introduced Earthworms." *Ecology* 67, no. 4 (1986).

Jordan, Diann, et al. "Earthworm Activity in No-Tillage and Conventional Tillage Systems in Missouri Soils." *Soil Biochemistry*, in press.

Jordan, Diann, John Stecker, and James Brown. "Earthworm Activity in No-Tillage and Conventional Tillage Systems in Missouri Claypan Soils." *Agronomy Miscellaneous Publication* 95-02, University of Missouri, 1994.

Reynolds, John. "Status of Exotic Earthworm Systematics and Biogeography in North America." In *Earthworm Ecology and Biogeography in North America*, ed. Paul F. Hendrix. Boca Raton, Fla.: Lewis Publishers, 1995.

Reynolds, John, and Wilma M. Reynolds. "Earthworms in Medicine." *American Journal of Nursing* 72, no. 7 (1972).

Wood, Hulton B., and Samuel W. James. *Native and Introduced Earthworms from Selected Chaparral, Woodland and Riparian Zones in Southern California*. General Technical Report PSW-GTR-142. Washington, D.C.: U.S. Department of Agriculture, U.S. Forest Service, 1993.

7

S EVERAL SUMMERS AGO, Arne and I spent a week in a
cottage on the back bay of Delaware's barrier island. The
moon was waxing, and horseshoe crabs were coming in to
nest. They are, I knew, considered primitive animals, yet they
had sensitivities I don't have, an exquisite alertness to the moon's
gravitational pull and an appreciation of the subtlest changes in
light levels. And those sensitivities let them live in a world of
mud, moon, and morning different from any world of those
parts that I could possibly know.

Anyone who has ever walked anywhere along the Atlantic
shore from Maine to Yucatán (or, for that matter, on the Pacific
shore from southern Japan to the Bay of Bengal) has probably
seen horseshoe crabs. A friend of mine, a Midwesterner, saw them
for the first time one day years ago when he and I were walking
along a stretch of beach in New Jersey. A dead horseshoe crab had
been washed up by the waves and lay, overturned, on its carapace,
its underparts, many legs, and folded, sheafed gills exposed. We
stopped to look at it. My friend said it reminded him of a fossil
he'd once seen in a museum, a trilobite, he remembered it was
called. He'd thought they were extinct.

A horseshoe crab does look something like a trilobite with a

long tail. Horseshoe crabs are big, sometimes two feet in length
from stem to stern. Right side up, they resemble, in their dun,
military color, oversized battle helmets or possibly tanks with a
sword sticking out. They live out most of their lives in the soft
mud of ocean bottoms, but during the warm months they come
up to the intertidal zone to nest and lay eggs, lumbering along,
looking impossibly clumsy. Their shape is so peculiar — that tank
with a sword business — that when they are standing still it is hard
for a person unfamiliar with them to figure out which is front and
which is back. Is the sword truly a sword for poking and impaling?
Or is it perhaps the mother of all stingers? I've seen parents on
beaches pulling curious children away from horseshoe crabs, so I
guess they must look fierce. They are not. They are about as mild
and inoffensive as any animal alive. That long spiky sword is something like a tail and is used as a stabilizer when they are swimming,
which they sometimes do on their backs. On land they walk along
on their legs, right side up, but if they get flipped by a wave or
other accident, they use that tail to lever themselves over.

My midwestern friend was astute in a way he did not know
when he said that the horseshoe crab we found reminded him of
the trilobite fossil in the museum, because the former are considered to be the closest living relatives of the latter. Although the lineage is far from clear, some ancestral form of horseshoe crab was
sharing muddy ocean bottoms with trilobites in the Cambrian,
more than 500 million years ago. The trilobites, after at least 300
million years of living in that mud, died out by the end of the Permian, while something quite like modern horseshoe crabs lived
on, along with other members of the subphylum into which they
are grouped, the Chelicerata. And by Jurassic times, 200 million
years ago, animals virtually identical to the one my friend and I
found on the New Jersey shore or the ones coming in to nest
in the Delaware bay were around and have been preserved in the
fossil record.

In addition to the horseshoe crabs (which are not true crabs — those are crustaceans), the Chelicerata include spiders, daddy longlegs, mites, and ticks. They all have two-part bodies: a head made up of fused segments and the other part, all in one piece. They all have four pairs of walking legs and some other modified legs, too, one pair of which are called chelicerae, from the Greek words *khele,* which means talon, and *keras,* which means horn. Those clawlike legs are, functionally, parts of the animal's mouth. They are used for mauling, crushing, and processing food before doing what we think of as eating, that is, putting it inside their bodies. But if you are equipped with chelicerae, the handling of food outside the body is part of eating, too. What horseshoe crabs crush in their chelicerae are marine worms and burrowing mollusks such as clams, but they are not above taking up other edible bits and pieces that have floated down through the water and settled alongside them in the mud.

Although the precise ancestry and phylogenetic line of horseshoe crabs are not absolutely clear, we know quite a lot about their biology and the way they live because they have been useful commercially and scientifically. Horseshoe crab research even led to a Nobel Prize for one researcher, H. R. Hartline, in 1967, for his work on the neurophysiology of vision. Horseshoe crabs have four eyes, two simple ones, and two compound ones, all on their backs. The latter interested Hartline because they are special in a number of ways. No other living chelicerates have compound eyes, although insects do and the extinct trilobites did. Compound eyes, unlike ours, are made up of bundles of small tubes, each receiving light at a slightly different angle. Animals that have eyes like that are very good at seeing movement but not very good at making whole, coherent pictures. Among the insects, those bundled tubes are so tightly packed that they are hard to study, but the horseshoe crab is a big animal, with big eyes in loosely packed bundles, so it is easier to figure out what is going on in the neural wiring of each tubule. That is what Hartline did, and he was able to apply his newfound understanding to the vision of other animals, even ours.

Horseshoe crabs have other attributes that make them a handy research animal for biologists. Woods Hole, the center for oceanographic research on the southern shore of Cape Cod, Massachusetts, is near the prime breeding grounds of *Limulus polyphemus,* the only species of horseshoe crab found along the Atlantic coast, and the work done on them there has become famous. During the 1950s, Frederick Bang, while working at Woods Hole, found out what made the blood of *L. polyphemus* clot. Horseshoe crab blood isn't red like ours, which takes its color from the iron used to carry oxygen. Theirs is blue, like the blood of pill bugs and of many mollusks and arthropods. It is blue because it carries oxygen not on iron but on copper. Horseshoe crabs are favored for study by blood researchers because they are big, easy to collect, and have well-developed circulatory systems in which the blood collects in two large pools. Best of all, at least for researchers, nearly one-third of the animal's total blood can be taken at a time without, they claim, hurting it. As a result of Bang's work on the clotting factor, a number of pharmacologically useful products were developed, including one that has been used routinely to assure the purity of drugs.

Before *L. polyphemus* became the white mice of Woods Hole, they were of interest to humans in other ways. In New England, where people like to eat clams as much as horseshoe crabs do, they were (and still are) regarded as competitors and therefore a nuisance. At one time a bounty was even offered for their killing. And ever since people have fished these coastal waters, horseshoe crabs have been cut up and used for bait. In the Orient, people eat them more directly, or at least eat their eggs. In the early part of the nineteenth century, entrepreneurs along the Delaware coast began to collect and process horseshoe crabs for animal feed and fertilizer. By the 1850s they were being "harvested" by the millions. The photograph below was taken in 1924, at Bowers Beach, not far from the cottage where Arne and I stayed years later. The picture shows horseshoe crabs stacked, awaiting processing at a nearby fertilizer factory. The numbers of animals taken were so great that by the 1950s the population had dwindled to the point that the industry was no longer profitable. But horseshoe crabs live many years and lay many eggs, so after the factories closed, the population recovered.

Although our North American *L. polyphemus* was named by Linnaeus, it was first described in 1588 by Thomas Herriot, mathematician, scientist, and sometime correspondent of Johannes Kepler. Herriot was tutor to Sir Walter Raleigh who, in 1585, appointed him as geographer to the second expedition to Virginia. It was there he first observed the animal he named horsefoot crab.

An important part of the life cycle of these shore-nesting animals is easy enough to observe and before long was traced out. In summer they come in from deeper waters to spawn during the high tides of the new or full moon. Many animals living

Horseshoe crabs near Bowers Beach, Delaware, June 17, 1924.
Courtesy of Delaware State Archives

near the shore have a thirteen-hour internal tidal clock and a twenty-seven-day lunar sensitivity that allows them to respond to the tides. Robert Barlow and others studied the mating cycle of *L. polyphemus* around Woods Hole, where the tides are unequal. They found that the animals preferred to nest during the higher of the two daily high tides. However, tides match not the solar but the shorter lunar cycle, so they progress. The animals would follow the higher tide through its changing phases into the dark hours, but as soon as that tide began to nudge the dawn, they would shift to the afternoon tide and follow it into the night. Barlow and his fellow researchers theorized that *L. polyphemus* took their clues from light.

After the horseshoe crabs have climbed the shore to the highest limit of the tide, the females begin to dig nests as the water recedes, laying several hundred to several thousand one-eighth-inch pale green eggs within the nests. The smaller males, clinging to the females' carapaces, fertilize the eggs with milt as soon as they are laid. As the moon pulls the high tide lower during the month, the developing larvae are safe from swimming predators, if not from gulls and other seabirds, who learn to peck them out of their nests. The next moon's cycle again draws water higher and higher until it washes over the nests, and the tiny horseshoe crabs who have escaped becoming gull dinner are rinsed from their nests by the tide and taken back into the sea. At this point they are called trilobite larvae because, with only the merest nubbin of a tail, they look very like those extinct animals. In the water they swim off, upside down, paddling with their gills, just as their elders do. Those fleshy gills that you can see when you flip a horseshoe crab on its back are called book gills because they are arranged in many leaflike folds. The arrangement is very like that of the lungs of their terrestrial cousins the spiders, which have correspondingly named book lungs.

In the sea the larvae begin the process of growing up, feeding,

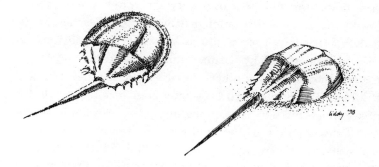

Left: A fossilized horseshoe crab. *Right:* A present-
day living horseshoe crab (1/12 lifesize).

growing a new carapace, and shedding the old one as they increase in size, perhaps sixteen times in all before they are mature. Most of their time is spent partially submerged in the mud, foraging for clams and worms. Their distinctive shape and lumbering gait, which make them look clumsy on land, are efficient in the mud. Males reach adulthood in their ninth year, females a year later. After that they follow the pull of the moon to the beaches for spawning for five or six years.

People like to call horseshoe crabs living fossils. They are considered primitive in the sense that the four species alive today are very like the ones found as fossils that lived millions of years ago. They show enormous conservatism in form and style of living over those years and few modern improvements. They are the only surviving members of a class of animals called Merostomata. Some members of that class, horseshoe crab cousins, the Eurypterida, were swimming about, along with the trilobites, during the early part of the Paleozoic. Like horseshoe crabs, the eurypterids were big animals, some of the biggest invertebrates known. They left behind spectacular fossils. One kind measures nearly ten feet in length. By the middle of the Paleozoic, in Silurian times, those millipede days, eurypterids were poking about in brackish lagoons and beginning to walk upon beaches, for they had gills that stayed moist, just as horseshoe crabs do today. But they became extinct by Permian times, around the same time as the trilobites and many other animals. Three big invertebrates — trilobites, eurypterids, and horseshoe crab ancestors — all living pretty much the same way, ploughing soft ocean muds, at pretty much the same time. One kind survived and produced descendants that live today, but two didn't. Why? Were the horseshoe crab ancestors better adapted somehow? Or were they simply lucky? The answer would depend on what kind of Darwinist you asked.

One hundred and fifty years of evidence has proved that Darwin had it right for the most part and biologists today are all basically Darwinists. To be sure, the evidence has shown that some aspects of his theory needed correcting. Darwin thought,

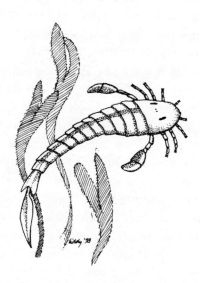

A eurypterid (¹⁄₁₀₀ lifesize), as it appeared in Silurian times.

for instance, that characteristics exhibited by parents might be blended in their offspring, but genetics, which he did not know about, has given us a crisper, more digital way of looking at heritability. But in the main, discoveries have shown his theory to be accurate. In addition to all the new evidence from genetics, there are new fields of knowledge — paleozoology, microbiology, population biology — which, based on a framework of Darwin's evolutionary understanding, have extended, expanded, and altered basic Darwinism. Today there are Darwinists who answer to a lot of different labels: neo-Darwinists, post-neo-Darwinists, new synthesizers, gradualists, neutralists, molecular drivers, mosaicists, and the punk eek guys — that is, advocates of punctuated equilibrium.

Niles Eldredge, curator of invertebrates at the American Museum of Natural History in New York City, is one of the punk eek guys, and his scientific specialty is trilobites. In 1970 he published a paper recording his research with horseshoe crabs, which he had kept captive in order to see how they burrowed. Because they are so similar in habit and body build to trilobites, he was investigating their burrowing to gain some understanding of the way trilobites might have lived and used their body parts, particularly their legs. His particular interest became speciation in trilobites. His colleague Stephen Jay Gould was studying speciation in land snails. Gould, also one of the punk eek guys, is perhaps more widely known through his lively and popular monthly column in *Natural History* magazine and his best-selling books. Both men are writers of graceful, accessible works about evolutionary biology and have given the reading public an understanding of Darwin and his thought. Punctuated equilibrium is one take on Darwinism. Its hypothesis states that after long periods of relative stability, evolution proceeds quickly (at least in geological time) through speciation at the edges of a population and is associated with catastrophe that extinguishes the main population.

Eldredge and Gould did not originate the basic hypothesis, but they did stitch it together in a new way, based on work by Ernst Mayr (the same man who had brought Klaus Ruetzler to the Museum of Comparative Zoology to elucidate the sponges there), to which they added observations from their own research. Darwin had pointed the way himself in admitting that according to his theory of gradual change, the fossil record ought to have turned up many "insensibly graded" variations, making a bridge between one species and another. That such variations had not been found he attributed to the fewness of known fossils, expressing the hope that as more were dug, transitional forms would be found. But in later editions of *The Origin of Species* he also admitted that "many species when once formed never undergo further change but become extinct without leaving modified descendants, and the periods during which the species have undergone modification, though long as measured in years, have probably been short in comparison with the periods during which they retain the same form."

Punctuated equilibrium holds that the lack of transitional fossils in the record is not a problem, that on the contrary it helps prove the hypothesis. A basic principle in modern science is that of falsifiability, which states that a hypothesis is acceptable after it has withstood all attempts to prove it false by experiment or contrary evidence. A corollary of this (and of interest when creationist ideas are brought out) is that propositions incapable of being tested, such as those of metaphysics or religion, are not falsifiable and therefore not science. Eldredge and Gould published their first paper on punctuated equilibrium in the early 1970s, and it has been put through the lively process of testing for falsifiability ever since. It has been pinched and prodded and measured and squeezed and weighed and held up to the light and knocked about a good deal in a passionate way. Scientists are a skeptical, contentious, argumentative lot. Not every scientist is yet convinced of the validity of the hypothesis by any means, but many are, and many others say that evidence supports it well enough to explain some, if not all, of the evolutionary process. So it is interesting to trace how at least one of the punk eek guys arrived at his conclusions.

As Niles Eldredge tells it, he was a young man in search of a doctoral thesis. He had chosen to work with trilobites partly because they are some of the commonest fossils to be found in sediments dating back to the Cambrian in this country (that is why they are so common in museums) and partly because they are found in a wide range from Michigan to Pennsylvania. He planned to study them throughout that range and to trace the changes as they speciated in the slow, stately, gradual shifts that Darwin, in the beginning, had hoped to find. What Eldredge discovered, however, was that throughout their 8–million-year history, trilobites changed so little that there was virtually no evolutionary trend at all — with one important exception. Around the edges of the range he observed "little side branches of new species splitting off from the main stock, and these descendent species living on side by side with the ancestral stock."

Each individual animal carries certain varieties of the genes that characterize the species; others of that species carry other varieties of the same genes. We humans, for instance, have genes that determine our eye color — some have genes for brown eyes, others blue, others hazel. In a big, homogeneous population the gene varieties get tumbled around and shared about steadily, and they show up in just about the same ratio all the time. But groups of individuals that live out in the suburbs of a population, being fewer in number, may not show the entire range of all the genes shown by those in the center, or they may possess them in a different ratio. As the suburbanites mate with one another and adapt to the special conditions in their neighborhood, they become more and more distinct genetically from the urbanites. At first they just do not mate with the main population or those from other suburbs, and then they *cannot* mate with the others — which is how a species is defined. And some of their now-distinct combinations of genes allow them to better live where they are, which is slightly different, perhaps in terrain or food, from where the others live. And so they not only manage to live in that particular spot, but they thrive there.

Such a situation is, I suspect, what is going on with the camel crickets I wrote about at the beginning of this book. Genes control a lot in addition to eye color. They also control biological processes such as rate of maturation and development of sexual capabilities. The camel crickets who lived in the territory around my farm appeared identical to the ones living in a circle beyond it, except that the males inside the circle grew a big orange bump on their backs just below their heads when they were mature and ready to mate. And they were randy and ready long before the camel crickets outside the circle were mature. In captivity I could induce the two groups to mate (and produce odd, monstrous offspring), but in the wild they never did because, maturing at different times, each mated with its own kind.

Now, if suddenly the globe warms enough to make the seas rise spectacularly and the waters rush into the central part of North America, as they have done, leaving only the high points, such as the one where my farm was, above water, the orange-bump camel cricket population would live on and the ones without bumps would all be drowned; they would be extinct. And when the waters receded, the orange-bump population would spread to cricketless habitat. Taking up all the special niches available to their kind, they might very possibly split up rather quickly into additional species.

My speculative story is somewhat analogous to what Eldredge found had happened to his trilobites in the fossil record. They did not become extinct all at once but were subjected to a series of local catastrophes. The ocean in which most of them had lived, covering the central part of North America, dried up extensively but not completely. And as it did, the trilobite species that had inhabited the area, *Phacops milleri,* died out. When the seas returned for a time, a species that had lived contemporaneously

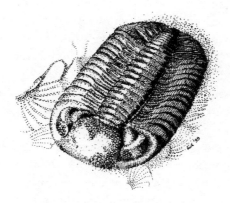

A trilobite fossil, *Phacops milleri.*

with *P. milleri* and continued to live around the edges where there still was some ocean water, *P. rana,* recolonized the renewed marine environment.

In the end, the crisis of Permian times, which, it is now thought, may have been related to an upwelling of carbon dioxide, extinguished all the remaining trilobites. But the horseshoe crab line survived. Eldredge and Gould lay more emphasis on chance than do some other evolutionary biologists, and they might say that the horseshoe crab ancestors survived when their cousins didn't because they were just lucky, not because they were better adapted.

When I learned my biology, and still up to twenty or twenty-five years ago, natural selection was seen as a process that turned out animals optimally adapted to their particular niche, engineered to fit. But most zoologists today see a looser, chancier, more haphazard process, in which natural selection acts more like a country carpenter, making do with whatever is at hand, rather than a master cabinetmaker using store-bought materials.

Take the trilobite's eye, for instance. An arthropod's eye is covered with a lens, a thickened layer of the animal's cuticle. In most arthropods the cuticle is made of chitin, a hydrocarbon allied to cellulose, and protein, sometimes reinforced by minerals. In trilobites, however, those who have studied them have deduced that the cuticle was not chitin but calcite, a mineral. Calcite is crystalline and, although clear, it has the pesky property of doubling images seen through it, not a helpful property for an eye lens, one would think. And yet natural selection fussed with the basic structure and rejiggered the physiology and came up with what is thought to be a very creditable eye, although different from others we know about. The trilobite eye, the oldest visual system known, was all jury-rigged from materials at hand. An engineer would have specified a non-image-doubling material right from the beginning.

Evolutionary history is littered with similar tales and what Stephen Jay Gould likes to call "spandrels," which is an architectural term referring to the space left over between a rounded arch and the straight lines of wall and ceiling. These biological spandrels are "nonadaptive parts and behaviors," according to Gould, who writes:

> In principle, spandrels define the major category of important evolutionary features that do not rise out of adaptations. Since organisms are complex and highly integrated entities, any adaptive change must automatically "throw off" a series of structural byproducts. . . . Such byproducts may later be co-opted for useful purposes, but they didn't arise as adaptations. Reading and writing are now highly adaptive for humans, but the mental machinery for these crucial capacities must have originated as spandrels that were co-opted later, for the brain reached its current size and conformation tens of thousands of years before any human invented reading or writing.

Last winter I heard Stephen Jay Gould speak in Washington. He is a big bear of a man who ignored the usual glad-to-be-here

niceties of public speakers. Instead he simply walked to the po-
dium talking, talking, talking, spinning a web of wonderful words
from what seemed to be unconnected data in a variety of fields.
But soon he made surprising and compelling connections. As he
wrapped up his lecture, he pointed out that the catastrophe that
had destroyed the very well adapted dinosaurs had spared — for
no apparent reason, just by chance — small mammals that seemed
not particularly good at anything except possibly staying out from
under the feet of dinosaurs. Once the dinosaurs were gone, he explained, those little mammals had a whole world to fill. Lucky for
us. They were our ancestors.

In the summary to their 1972 paper, "Punctuated Equilibria: An Alternative to Phyletic Gradualism," Eldredge and Gould
wrote, "The norm for a species, or by extension, a community, is
stability. Speciation is a rare and difficult event that punctuates a
system in homeostatic equilibrium . . . an uncommon event that
produced . . . a wondrous array of living and fossil forms."

Research since that time has produced a number of examples of
such rapid evolution after stasis. But still other research, even with
trilobites, has shown examples of gradual, steady speciation, so
those who are cautious say that all we can claim with assurance is
that the pace of evolution can vary. Research in molecular genetics
is beginning to reveal the complexity of the way genes work. The
increased understanding of such processes expands and refines
Darwinism but shows that 150 years ago the man had it basically
right: that forms of life on this planet come from other forms of
life, not from supernatural agency; that descent with modification
occurs through the force of natural selection.

Darwin's theory was enormously controversial when he published it, and all this time later there are still people who find the
thought of evolution, driven by any mechanism, gradual or in
rapid bursts of speciation, unsettling. I know schoolteachers from
Alaska to Virginia who have been harassed by a noisy minority to

withdraw school textbooks, alter lesson plans, and amend their lectures when they mention evolution. One hundred and fifty years after Darwin, seventy-five years after Clarence Darrow and the Scopes trial, it is bizarre that we still have teachers who are afraid to talk about evolution in the classroom, even though they can give no proper instruction in biology without it. This was made clear to me one evening a couple of years ago in Missouri when I was having dinner with friends and their daughter, a bright, vivacious, intellectually sophisticated high school sophomore. Both of her parents are professionals, although not scientists. We were lamenting a recent scheme in a nearby town by creationists to unseat the local school board.

"I don't see what they are so upset about," said the sophomore. "After all, evolution is just a *theory;* nobody is even saying it's true."

I explained to her that a theory, as the term is used by science, is a broad intellectual framework for observed events; that in science, unlike religion, nothing is ever said to be "true" but is always subject to emendation by evidence and experiment; that science is a process, not received wisdom. The theory of evolution, I told her, is as valid as atomic theory, cell theory, or the theory of relativity, all long substantiated by evidence. She nodded. She knew about those theories, didn't think there was any problem about them. She seemed interested but astounded that the theory of evolution was as valid as those others, reflecting, I'll wager, her biology teacher's hedging presentation to avoid the wrath of creationists and their agenda.

But what of the horseshoe crabs, those dun beasts who helped Eldredge elucidate trilobite burrowing behavior? Just now, while we've been wandering through the human construct of evolutionary thought about what they and their kind might or might not have been doing over the millennia, they have been following the moon's pull toward beaches (it is summer as I write, and the moon is once again waxing). Over those millennia their ancestors endured cataclysms that killed their neighbors, shifting sea levels, and changing earth chemistry; populations of their own recent forebears have endured being ground up into animal feed, pulverized into fertilizer, drained of their blood for our benefit. Enduringness, hundreds of millions of years of it, is what they are good at; it may be their specialty. They seem to represent the equilibrium part of punctuated equilibrium; their ancestors may have represented the punctuation.

Today biologists who study horseshoe crabs report unusual massive die-offs as human activity changes the composition of their ocean mud homes. Development takes away their beach nesting spots. I've never heard of an organized movement Save the Horseshoe Crabs, however. Their placid usefulness, their assumed large numbers, their clumsiness, their resemblance to military hardware, all combine to keep them from seeming adorable. Even Linnaeus, who was an unpleasant man and often made mean jokes in his namings, gave our Atlantic coast species a derogatory name, *L. polyphemus.* Polyphemus (the Greek word means "much spoken of, famous"), it will be remembered, was the single-eyed giant blinded by Odysseus. But before Homer told his tale, Polyphemus was already a figure of fun and ignominy in folklore. He was a big, awkward, bumbling shepherd with a vile temper who wooed Galatea, a nereid, and lost her to another suitor. In Linnaeus's time he was known to the public as the big, dumb bumpkin in the 1720 dramatic cantata by Handel, *Acis and Galatea.* Linnaeus, in naming the animal for Polyphemus, was mocking its clumsiness as it lumbered out of the water to nest.

But there is another version of the legend that Handel did not use. In it, Polyphemus woos and wins the dainty sea nymph as his wife and lives happily ever after. I wouldn't count our *L. Polyphemus* out for another eon or two.

READINGS

Barlow, Robert B., Jr., et al. "Migration of *Limulus* for Mating: Relation to Lunar Phase, Tide Height, and Sunlight." *Biological Bulletin* 171 (October 1986).

Bonaventura, Joseph, et al., eds. *Physiology and Biology of Horseshoe Crabs.* New York: A. R. Liss, 1982.

Campbell, Neil A. "A Conversation with Niles Eldredge." *American Biology Teacher* 52, no. 5 (May 1990).

Clarkson, E. N. K. *Invertebrate Palaeontology and Evolution.* London: Chapman and Hall, 1993.

Cohen, Elias, ed. *Biomedical Applications of the Horseshoe Crab (Limulidae).* Proceedings of a symposium held at the Woods Hole Marine Biological Laboratory. New York: A. R. Liss, 1979.

Darwin, Charles. *The Origin of Species by Means of Natural Selection.* 1859. Reprint, New York: Humboldt Library of Popular Science Literature, 1884.

Eldredge, Niles. *Fossils: The Evolution and Extinction of Species.* New York: Abrams, 1991.

————. "Observations on Burrowing Behavior in *Limulus polyphemus.*" *American Museum Novitates,* no. 2436 (November 4, 1970).

Eldredge, Niles, and Stephen Jay Gould. "Punctuated Equilibria: An Alternative to Phyletic Gradualism." In *Models in Paleobiology,* ed. Thomas J. M. Schopf. San Francisco: Freeman, Cooper, 1972.

Gould, Stephen Jay. "Evolution: The Pleasures of Pluralism." *New York Review,* June 26, 1997.

Novitsky, Thomas J. "Discovery to Commercialization: The Blood of the Horseshoe Crab." *Oceans* 27, no. 1 (1984).

Stormer, Leif. "Arthropod Invasion of Land During Late Silurian and Devonian Times." *Science* 197 (September 1977).

8

WE HUMANS ARE a fussy lot. If the world isn't quite to our liking, we poke it and prod it and rearrange bits of it. Fiddling with the world is what humans do; it is a definition of humanness. We try to make it convenient, cozy, and pleasurable to ourselves. And so, when the first European settlers came to these shores in the seventeenth century and discovered, to their dismay, that there were no honeybees here, they set about to import them.

Never mind that more than 3,500 species of native bees were buzzing about, living their own lives, and energetically pollinating the native wild plants and the few crops grown by the people the settlers called Indians. Those bees produced neither wax nor honey. Only honeybees do that, and without them the Europeans lacked beeswax to make candles to light their dark days and darker nights; they had to make do with bayberry or wax myrtle wax or tallow from beef and deer. Without honeybees they lacked the major ingredient of mead, which is fermented honey. And they lacked honey to sweeten their food, although this was less important than it might seem, considering that honey was the chief sweetener in those beginning days; they did not have molasses until later. Eva Crane, the world's foremost authority on the history of honey and its uses by mankind, has studied English diets of the period and has concluded that they were far less sweet than ours are today. She contends that today's taste for sweet food is the result of the relatively recent availability of cheap processed sugar.

We have a couple of tantalizing hints of various species of small, social, stingless honey-making bees native to the warmer parts of the Western Hemisphere. They are distant cousins to the European honeybee, *Apis mellifera*. One report of honey, or at least "a variety" of honey, in what is now Cuba was reported by Christopher Columbus in 1492, but he did not say that bees had made it. Another is found in the journal of Hernando de Soto's marauding passage through the Southeast in 1539–41, as translated by Richard Hakluyt in 1611. The de Sotans paused long enough from murder, rapine, pillage, and conquest to note that in a "towne . . . between two armes of a river" in what is now Georgia, they found "butter, mealted like oile . . . the fat of bears . . . also a great store of oile of walnuts . . . of good taste, and a pot full of honie of bees, which neither before nor afterward was seene in all of the countrie." If this report is accurate, the probable maker of that "honie" would have been stingless bees, which might have been able to live in the warmest parts of Georgia.

A few years ago, while I was committing journalism in Guatemala, I visited a Mayan family that kept stingless bees in tiny, brightly painted hives about the size of shoeboxes. They kept the bees, as had their ancestors from Classical times, for their honey and wax. Those stingless bees are no longer found in any part of North America, although in my twenty-five years of beekeeping I've heard it proposed now and again that we ought to import them because it would be really nifty to have bees that don't sting. It is true that they lack stingers, but I can testify that they are very fierce about pulling the hair of anyone they consider an intruder, and some species are called in Spanish "fire shitters" for their ability to eject a nasty burning liquid chemical when disturbed. But they are bees of warm places and probably didn't thrive as far north as Georgia, let alone New England. Also, they are little bees and produce small amounts of wax and honey. They would not have been of interest to the North American colonials even if the settlers had known about them.

Those early settlers weren't interested in the pollination work that the native bees — or even honeybees — do because they didn't know about it; the role of insects in plant fertilization wasn't clearly understood until the late 1700s. That was long after the first known importation of European honeybees to this continent, which is recorded in a letter dated 1621, written from the English office of the Virginia Company. It says, "Wee have by this shippe sent . . . fruit trees, as also pigeons . . . and bee hives . . . the preservation and increase hereof wee recommend unto you." Virginia would have been a happy landing for honeybees, which do well in a temperate climate, even though they, like their stingless cousins, probably originated in a tropical one. If they survived the ocean voyage, they would have prospered at that latitude and may have been the origin of cast swarms that in following years were noted spreading westward. Although an exact date has not been pinpointed for New England, sometime between 1630 and 1663 honeybees were also imported there and lived to brighten the lives of the colonials.

By the mid-eighteenth century, the bees had crossed the Appalachians and were found wild in trees or in hives made from logs (these were called gums, because they were made from gum trees, which often hollow as they age), at pioneer homesteads. Fifty years later, by century's end, they had crossed the Mississippi and were swarming, feral and free, nesting in tree hollows up the Missouri and other western rivers. There is some disagreement about the date, but early in the 1800s, honeybees were imported by ship to the West Coast, and their population was soon increased by hives brought overland by settlers. Within the next half century,

with the invention of better beekeeping equipment and an industry to manufacture and sell it, honeybees were to be found, feral or
cared for, throughout the United States. In roughly two hundred
years an exotic — that is, imported — insect, the most important
(and often only) pollinator of the exotic crops that white North
Americans wanted to grow, had spread from coast to coast. It had
become a necessary, if unacknowledged, partner to American
farming.

The invasion of exotics was quickly noted by the Indians, who
created names for them: white man's weed, the daisy; white man's
foot, the plantain; English man's fly, the honeybee. The bees in
particular, because of their habit of swarming out of their cramped
log gums and preceding — and thus predicting — the arrival of
whites, were regarded with a certain amount of sourness, it is said,
and became the equivalent of our sighting of a black cat.

The Indians who were farming grew corn, squash, and beans
(also exotics from the neotropics). But corn doesn't need insect
pollination, and the native bees were adequate for the pollination
of small plots of squash and beans. The white man's farming, however, was based on a whole range of imported plants, many of
which needed bee pollination but bloomed at times other than
those of the population peaks of native bees. In addition, the
native bees were unruly and could not be managed or put exactly where the crops were. So by the late 1800s those interested
in farming were beginning to recommend setting out honeybee
hives in orchards to increase fruit yield through pollination.

It wasn't until the middle of the present century, as agriculture became agribusiness, that honeybee pollination services were
turned into business as well. With the shortage of men to do heavy
hand labor during World War II, and with the development of
farm machinery, farms became bigger and machines took over the
hand labor. Large plantings of single crops — monoculture —
became the rule, and chemical fertilizers and herbicides were re-

quired to grow them profitably. Many of the native bees, even if they are able to pollinate the exotic crops, are small and have restricted flight ranges: most fly no more than one hundred to two hundred yards. They would not penetrate very far into big fields and could not nest within them, because their usual nesting spots, in hedgerows and trashy, weedy places, are destroyed by "clean" cultivation. In addition — and this always seems to surprise farmers when I mention it — bees, native or honey, are insects and are killed by insecticidal sprays. People seem to want to believe that insecticides kill only bugs that we have declared "bad."

As a result of all of these changes in agriculture, honeybees have become more important. People who keep them have learned to take advantage of both mechanization and the farmers' need for pollination. They group four hives on a pallet, use fork-lift trucks or boom loaders to load and unload the pallets of them from trailers, and truck the hives from crop to crop, moving them in for pollination when the flowers are blooming, then moving them out again before insecticides are applied. For this they receive a rental fee that has become, for some beekeepers, a more important source of income than the honey the bees may produce. In recent years another exotic bee — the Eurasian alfalfa leafcutting bee, which can also be managed easily — has also been used for pollination. Together these two kinds of bees account for 90 percent of the crop pollination in this country, according to Suzanne Batra, a research entomologist at the United States Department of Agriculture. She also estimates that about one-third of our food supply depends on insect pollination.

In 1933, for the first (and, one assumes last) time, John Kenneth Galbraith was the junior author of a pamphlet entitled *Economic Aspects of the Bee Industry,* which indicated that for the previous seven years the hive rental fee for fruit and almond trees in California had been two dollars, tops. Today, as the century ends, orchardists in that state are paying twenty times that fee, and in other places, such as here in Maine, where "wild" blueberries are big business, even thirty times. Inflation, of course, plays a role in that increase, but it is a minor one. The real reason for the high price is that demand exceeds supply, because all across the continent honeybees are in trouble. The press has been running stories about the "pollination crisis" caused by the sudden recent shrinking of honeybee populations because of parasitic mites and disease. But for beekeepers willing to cosset and medicate their bees, for the beekeeping industrialists manufacturing the products to do so, and for the bee breeders with capacity to raise new stocks to replenish dead hives, the "pollination crisis" is an opportunity and good business.

Honeybees, like other animals, get sick and have parasites, but over the years beekeepers in this country have learned how to help their bees stay healthy. However, when a new disease or parasite appears that the host animal has no evolved defenses for, it can spread rapidly, particularly among animals that are kept in unnaturally large populations, as honeybees have been. For instance, leafcutting bees kept in sufficient numbers for alfalfa pollination become attractive victims to a parasitic wasp.

It was 1904 when parasitic mites, *Varroa jacobsoni,* were first noted on honeybees in Indonesia. Those bees are a different species — *A. cerana,* not *A. mellifera* — but like our bees, they nest in dark places, so they, too, can be kept in hives. Beekeepers all over the Far East, in Japan, China, and India as well as Indonesia, keep *A. cerana* and use them to produce honey and pollinate crops. *A. cerana* and *V. jacobsoni,* which parasitizes the bees' larvae, have evidently lived together for a very long time, for the mite is not a serious pest in the Far East. Mites are found on the developing young bees, to be sure, but not in great numbers. The adult bees pick the mites off the larvae and groom one another to take off and bite to death any that they find. *A. cerana* has also developed tolerance of several other kinds of mites through long association.

Over the years, *V. jacobsoni* spread westward, through the former Soviet Union and then to Europe, where beekeepers had
their hands full already learning to deal with another parasitic bee
mite, this one named *Acarapis woodi*. That mite parasitizes adult
honeybees, crawling into their tracheae and slowly killing them. It
was just a matter of time before both mites showed up on our
shores. As a beekeeper, I have always been disappointed that our
for-profit bee breeders, government-funded researchers, and the
USDA did not begin working on developing resistance in our
honeybees earlier in the century, when *A. woodi* had already become a problem in Europe. Instead we put up an imaginary barrier: in 1922 Congress passed the Honey Bee Act, which severely
regulated the importation of bees. But barriers never work. Like it
or not, we live in one world, a world in which human-facilitated
travel grows easier by the year and national boundaries harder to
define and defend, particularly against very small animals.

In the late 1980s I began losing more and more hives over the
winter, which is a trying time for bees that may be stressed by parasites, and soon I was noticing masses of them crawling on the
ground, a sure symptom of the energy-sapping tracheal mites, *A.
woodi*. In 1988, *V. jacobsoni* was first discovered in several American states. Another imaginary barrier was put in place: the bees
from those states were quarantined. Quarantines never work either, for flying creatures that become feral in great numbers. The
next defense against mites of both kinds was chemical treatments.
These are still sold, but evidence is accumulating that the mites are
becoming resistant to the chemicals. It is only in the past few years
that American researchers have begun to devote serious attention
to breeding mite-resistant bees that also produce honey. The results of their research are still years away.

Both mites spread rapidly across the country and today are
found in all parts of it. I started seeing the larvae-parasitizing *V.
jacobsoni* on my bees in the Ozarks a few years after I saw signs of

the tracheal mites, but by then I was no longer keeping bees on a commercial basis. I had reduced my stock from three hundred hives to just a few, and I could afford to hold out against chemical treatments. Instead, I tried to develop a resistant strain of them myself. It wasn't until 1996 that I was able finally to keep a hive of bees alive through the winter (something that in earlier years had been so easy) and bring it into springtime as a modestly productive hive, indicating that the untreated bees within were resistant to the two species of mites.

By that time few hives of kept bees survived anywhere in the Ozarks, and apparently there were no longer any feral honeybees at all. Fifteen years earlier it had been a rule of thumb around the country that there were approximately the same number of feral honeybees, living in tree holes and under the siding of abandoned buildings, as hived ones. Those feral bees did a lot of pollination work for free.

In Missouri, where farming is largely a matter of raising cattle, a nice stand of white clover in a pasture is appreciated. There are also some orchards in the state. But in recent years farmers were telling me that their white clover and their fruit trees were the poorer for lack of honeybees. In talking to other beekeepers and researchers around the country, I heard the same story. Many beekeepers, discouraged by the expense of replenishing their stocks of bees each spring and the growing uselessness of antimite treatments, have left the business. In some years when the weather was bad, 80 percent of the bee colonies alive at the beginning of spring did not make it through the next winter.

Researchers are beginning to take an interest in both our native wild bees and a handful of species imported as crop pollinators. Now there is a fledgling industry in these bees, called pollen bees to distinguish them from honeybees, for they do not produce any harvestable honey.

Unlike the social honeybees, which nest in large colonies and divide themselves up into castes, each with a different job to do, most native North American pollen bees are solitary. Each female bee makes her own nest, with about ten brood cells, in a hole in the ground or, depending on the species, in a stem, post, or rotting tree. She builds her nest cell by cell, provisioning each one with a mixture of nectar and pollen, laying a single egg in it, then sealing it off before moving on to make the next one.

Solitary bees range widely in size: the smallest North American species, *Perdita minima,* is about the size of a fruit fly; tropical carpenter bees may reach an inch and a half in body length. Many are generalists, but some are specialists: squash bees work American native gourds; a kind of miner bee prefers wild morning glories; an oil bee loves the oil of wild yellow loosestrife; and some night-flying sweat bees pollinate evening primroses.

I asked Charles Michener, of the University of Kansas, the grand old man of American bee and pollen studies, how native pollen bees have been affected by human activities and by competition with the honeybees. "Well," he said, "disturbed land and bare ground created by humans are better for wild bees than undisturbed, wooded land. Bees like open, sunny places." In 1983 some of his students compared disturbed land near Lawrence, Kansas, with pristine areas nearby and found not only more individual bees on the disturbed land, but also more species of bees. And an unpublished 1972 study by the Illinois Natural History Survey found, in an area near Carlinville, Illinois, the same 220 species of bees that had been there in the early 1900s. Michener cautioned, however, that the original data had been collected long after the introduction of honeybees and, since there is no census of numbers and diversity of species before the Europeans came, it is nearly impossible to assess the effects that we and our honeybees have had on native bees. "Very little work on this important matter has been done, except in Australia," he said.

David C. Paton, who has summarized the Australian research,

writes, "These studies show that honeybees may displace native pollinators from flowers and may not trigger the pollination mechanisms of the flowers they visit." But by pollinators he does not necessarily mean bees. He points out that the effects of honeybees since their introduction to Australia 150 years ago are complex and hard to quantify. Their effect on the native wild bees, at least on one level, does not appear harmful. The smaller wild bees visit more flowers more often, making up for the fact that the bigger honeybees are able to nudge them aside at the best nectar sources. It is with the birds that the story is different. Honeyeaters, Australian birds with long slender bills, are important and efficient pollinators there. Paton found that they visited flowers less frequently when honeybees were active and tended to concentrate on flowers least used by honeybees. The honeyeaters' reaction is understandable to anyone who has ever watched hummingbirds at one of those red plastic feeders and seen them back off with every sign of disgust when they found wasps and bees trying to take sugar water from it.

David Roubik is a scientist with the Smithsonian's Tropical Research Institute in Panama. "I've studied bees in Mexico, Panama, and French Guiana," he told me, "and have good data on the population of native bees for seventeen years." The situation is complicated in the tropics by the presence of the native stingless bees, which produce and store honey. Roubik suggests that the exotic honeybees may be able to outcompete the native ones because they forage over greater distances and, if necessary, move their colonies by swarming. "It is interesting to note that there are fewer bee genera in the Old World, where honeybees have been for a long time, than in the New World, where they have been for only a short time," he said.

Suzanne Batra, the USDA bee scientist, offered a different explanation: "I think the different number of genera in the two places reflects past climatic differences. While it is reasonable to

think that exotic bees will compete for pollen and nectar with native species, there is as yet no hard data, and I am not convinced
that competition is as important as some people think it is."

I visited Suzanne, a vibrant woman who wears her hair in a single long braid, at her office in Beltsville, Maryland. A field biologist, she was impatient at being held inside on a rainy day. Her
room was a zestful clutter of files, magazines, technical papers, car-
toons, and photographs. Stacked papers threatened to swamp the
postage-stamp-sized open space on her desk. From one stack she
pulled out a diagram from a population study on a wide variety of
species of East Coast pollen bees and showed me that their numbers are high very early in the spring, when they emerge, mate,
provision their nests, lay their eggs, and die. The adult population
plunges by late May, just as the dense forest canopy leafs out, most
forest flowers finish blooming, and honeybee numbers are building. These data show that different species do not forage on the
same flowers. Currently Suzanne is studying a species of ground-
nesting *Andrena* bees that are important pollinators of red maples,
pear trees, and early-blooming wildflowers. These native bees are
so tolerant of the cold that they emerge from their nests as the
snow melts in February, long before other bees are able to fly.

Further evidence that native bees can hold their own against
honeybees comes from random samples taken by Suzanne Batra
and Edward Barrows, of Georgetown University, at different
seasons in some West Virginia woods in 1991. The year's catch
turned up only 34 honeybees and 1,701 native pollen bees. The
proportions were similar in the years that followed. By 1991,
hived and feral honeybees were already shrinking in numbers because of the mites, so the beekeeper in me suggested that a few
years earlier there might have been more honeybees in the sample.
But the beekeeper in me has to stand corrected. Recently Suzanne
has also taken random samples near her Maryland laboratory,
where there are many mite-free, medicated honeybee hives that

are part of USDA's research on mites. Even there, of the bees caught in sweeps around fruit trees and blooming wildflowers, only about 10 percent were honeybees. The rest were all pollen bees.

Suzanne pointed out, however, that native bees may be more vulnerable to competition where honeybees are more abundant. One experimental study in Arizona, conducted by ecologist William Schaffer and entomologist Stephen Buchmann, both of the University of Arizona, showed that abundant honeybees do seem to drive away native pollinators of shindagger, *Agave schottii,* a native species of century plant.

Until more data are in, Suzanne is not convinced that honeybees compete seriously with native bees. She also agrees with Michener that humans have made good habitat for native bees by clearing land and opening up the dense forests that once covered much of eastern and central North America. In addition, crops and ornamental plants in clearings and meadows have given the bees new sources of food. But, she added, some large-scale agribusiness practices, such as use of pesticides, clean tillage, and, above all, huge acreage devoted to monoculture, may be harmful to native bees. "Some species have benefited from us; some have not," she concluded.

Suzanne hopes that the more researchers and farmers learn about pollen bees, both native and imported, the more they will want to employ them, and that with greater commercial use may come some changes in agricultural practices that will be beneficial to the bees. She showed me the plastic boxes now being sold for keeping a familiar pollen bee — the bumblebee, which is said to outpollinate honeybees ten to one. For some crops, such as red clover, bumblebees are the most important pollinator. The nectar-bearing part of the red clover blossom is too deep for the honeybee's tongue; the bigger bumblebees have longer tongues and happily apply themselves to the blooms. But problems need to be solved before bumblebee pollination can become commercially economical outside of greenhouses (in Holland they are used to pollinate those handsome all-season tomatoes still attached to the vine that we are beginning to see in supermarkets everywhere). Bumblebee colonies do not last from year to year, they are tricky to raise artificially, and, most important, they are expensive. Prices are coming down, but the last time I checked, in 1996, it cost nearly three hundred dollars to buy a box of ten to one hundred bumblebees, none of whom would be around for the next pollination season.

At last the rain stopped, and Suzanne took me outside to show me a few of the other species that seem to have promise as agricultural pollinators. Some, like the shaggy fuzzyfoot bees nesting in dry adobe blocks near her office, are exotics. About the size of honeybees but fatter and darker, these Japanese bees, *Anthopora pilipes villosula,* are lively, fast flyers even on that cool, gloomy day, which was too inclement for honey and native bees. Nearby we looked at nest setups for some other species: drilled slabs of wood, looking like giant cribbage boards, for alfalfa leafcutter bees; bent cardboard tubes for a variety of mason bees. Then we walked down a woodland path to see the ground nests of three species of what Suzanne has named polyester bees. To keep their brood cells from getting damp, these native bees, nesting in the sandy soil beside the path, waterproof their cells with a thin transparent film, made of a natural polyester, which they secrete from a gland on their abdomens. The weather was keeping them inside their closely grouped nests, but I could see that a few had been at work carrying grains of sand to the surface and arranging them around their holes. General tidying. When we returned to her office, Suzanne showed me a liner she had excavated from a polyester bee cell. With an extra flap that the female uses to close the cell after she has laid her egg, the clear liner looked for all the world like a tiny Baggie for a miniature sandwich.

The polyester and other native ground-nesting bees contribute importantly to the pollination of wild plants and of crops grown on small farms and gardens where insecticides are not used, but they are difficult to manage on the scale needed by agribusiness. They can seldom be attracted to artificial nesting spots, and it would require a great deal of hand labor to go about with a shovel digging up their nests. Some growers, however, have managed to establish beds of special soil that is acceptable to the ground-nesting alkali bee, which is sometimes used for alfalfa pollination in the West. Their success has led to a familiar problem: unnaturally dense populations of the alkali bees are especially vulnerable to their natural enemies, which include fungi, other microorganisms, and parasitic insects.

Partly because of the difficulties of domesticating our native solitary bees, Suzanne chafes against regulations that restrict her experiments with manageable imported pollen bees. Some scientists, worried about the possible negative effects on native species of yet more introduced bees, would like to see research limited to the natives. They think that the mites have brought good riddance to the honeybee. Some passionate nonscientist "greens," who have just discovered that honeybees are exotics, hold them somehow at fault for the so-called pollination crisis. One told me she wished all the honeybees would die out so we could go back to the way things used to be before they came here. But our continent today is different from the one the Europeans found when they began coming in the seventeenth century. It is full of different plants and crops, and a different life prevails.

Suzanne Batra sees her job as developing a stable of bees that are inexpensive and easy to manage, and she suspects that some of the best candidates may come from other countries, where our major agricultural crops come from as well. "Some people have a prejudice against exotics," she said, adding bluntly, "It is an anti-immigrant feeling. But after all, most of our crops, livestock, and we ourselves are exotics."

This whole question of exoticism is one, I find, on which people have hot opinions that become, in the end, political. I grew up in the 1930s, the daughter of a landscape architect in Michigan who was trained formally to use many exotic plants in his designs. But as he grew older and more experienced, he found many exotics to be unhardy and fussy in their needs, and he became noted for using native plants in his landscape designs. So from my formative years onward, I've been predisposed to nativism, which is currently a fashionable point of view. But nativism these days labels anything exotic as Bad, anything native as Good. Reductio ad absurdum. When Hitler came to power in Germany, he forbade landscape architects to use foreign plants in order to preserve Aryan purity. That would be funny if it didn't have such obvious racial overtones. And, ultimately, if we were to magic away all "exotics," we'd find ourselves and the rest of humankind huddling in a mass over the Rift Valley.

Even if we don't go that far, "native" is hard to define. Is it only those natives that are found in a particular field, a county, a state? Is a species of plant or animal that is native to North Carolina exotic in South Carolina? Is it that which was here before Europeans came? What about those plants brought by Indians when they migrated here? What about the spread of animals from South America into North America once the land bridge was formed? What about plants spread by airborne seeds? When nativists argue, they like to point out kudzu, English sparrows, and gypsy moths, exotics that have caused humans many problems. But apples, the basis of our most patriotic, nationalistic pie, are exotics. Who could hate an apple tree? Most of our food crops are exotics, and it would be hard to put a meal on the table without them.

One of the better days I've lived was one I spent with a man who knew more about ladybugs than anyone else in the world, Ken Hagen, who, I am sorry to say, died last year. At the time of my visit he was working for the Beneficial Insects Lab at the University of California at Berkeley, and he was interested in developing natural controls of pest insects in order to avoid the overuse of chemicals in agriculture. He told me that biological pest control with ladybugs had begun in California in the early 1870s with the importation of an exotic ladybug from Australia to control the cottony-cushion scale insect, which had become a serious pest in fruit orchards. That imported ladybug continues to thrive and keep the scale insects under control without insecticides. No harm there.

I'd asked Sam James about exotic earthworms and he, it will be remembered, had given me some good quantitative data about their harmful effects. Still, no one is recommending a roundup and deportation of nightcrawlers. They are here to stay. With honey-bees the question of harmful effect is less clear. After talking with scientists and reading the research papers, I have come to the opin-ion that we may have a problem with the agribusiness style of farming, we may have a problem with consumer expectations in the supermarket created by marketing, but those problems are not the fault of the honeybee. Those are human problems, and here I find myself parting company with the nativists and their national-ism. Some exotics are bad for humans. Some are good. Many, both bad and good, are here to stay, and we must learn their ways and figure out how to get along with them.

When I moved from Missouri, I gave my one mite-resistant hive to a friend who wanted to get started with bees. I sold my re-maining commercial equipment. But I kept back a few hive parts and tools. I'm counting on researchers to develop a mite-resistant honeybee that will not have to be doused in chemicals in order to thrive. I hope it will come before I am too tottery to keep bees in Maine. I miss them.

READINGS

Barth, Freidrich. *Insects and Flowers.* Princeton: Princeton University Press, 1985.
Batra, Suzanne. "Evolution of 'Eusocial' and the Origin of 'Pollen Bees.'" *Mary-land Naturalist* 39, nos. 1–2 (1995).

————. "Solitary Bees." *Scientific American* 250, no. 2 (1984).

Buchmann, Stephen L., and Gary Paul Nabhan. *The Forgotten Pollinators.* Washington, D.C.: Island Press, 1996.

Burroughs, John. "An Idyl of the Honey Bee." In *Birds and Bees.* Boston: Houghton Mifflin, 1879.

Crane, Eva. *Honey.* London: Heinemann, 1976.

De Soto, Fernando. *The Discovery and Conquest of Terra Florida.* Trans. Richard Hakluyt. 1611. Reprint: Burt Franklin, n.d.

Free, John. *Bees and Mankind.* London: Allen & Unwin, 1982.

Hive and the Honey Bee, The. Hamilton, Ill.: Dadant, 1975.

Maeta, Yasuo. "Utilization of Wild Bees." *Farming Japan* 24–6 (1990).

Michener, Charles, et al. *Bee Genera of North and Central America.* Washington: Smithsonian Institution Press, 1994.

Murray, Lee. "First Bees in Massachusetts." *American Bee Journal* (July 1976).

Oertel, Everett. "Bicentennial Bees." Pts. I–III *American Bee Journal* (February, March, April, 1976).

Paton, David. "Honeybees in the Australian Environment." *BioScience* 43, no. 2 (1993).

Roubik, David. *Ecology and Natural History of Tropical Bees.* Cambridge, Eng.: Cambridge University Press, 1989.

Voorhies, Edwin C., Frank E. Todd, and J. K. Galbraith. *Economic Aspects of the Bee Industry.* Bulletin 555. Berkeley, Calif.: University of California, College of Agriculture, Agricultural Experiment Station, 1933.

Rae, Noel, ed. *Witnessing America: The Library of Congress Book of First-Hand Accounts of Life in America, 1600–1900.* Stonesong Press, Washington, D.C., 1996.

Wolschke-Bulmahn, Joachim. "The Mania for Native Plants in Germany." In *Concrete Jungle,* ed. Alexis Rockman and Mark Dion. Juno Books, 1997.

9

HALF A MILE upriver from Blue Creek village, a Mayan Indian town of thirty-seven families in southern Belize, is a 200-acre nature preserve, managed by a North American herpetologist. The trees are cleared and the land is under cultivation right up to the edge of the preserve. But within it is some of the most strikingly beautiful rain forest in all of Central America.

I've been to Central America a number of times. At the beginning, I found rain forests overwhelming, incomprehensibly complex, deceptive, humbling. To a northerner they are foreign in the extreme. My first visit to a rain forest was in Guatemala, and I was awakened on the first night by a strange noise. I was puzzled and made faintly anxious by it. I couldn't identify whether it was animal, vegetable, or mineral in origin. Perhaps the lodge's electric generator had been turned on, I told myself. But the sound advanced, grew closer, surrounded my cabin. It no longer seemed mechanical; it sounded like the roar of approaching surf, although I was far from the ocean. It made the hair stand up on the back of my neck. It wasn't until the troop of monkeys was directly overhead that I remembered, oh, yes, howler monkeys. I'd read about their calls, but no description of them measured up to the reality.

Yet at the same time the rain forest is familiar in a bizarre, unsettling way. In that same Guatemalan rain forest, surrounding the excavated Classical Mayan site of Tikal, there were, common as pigeons in New York City, turkeys of the same size and habit as the wily and elusive brown ones that strut and gobble in the Ozarks. They were different in only one respect: they were a brilliant, shimmering, iridescent blue, startling and lovely to my gringa eyes. But they weren't unusual to the Indian boys who sold soft drinks and snacks to the Tikal visitors. In fact, the birds were so tame that they were something of a nuisance. The young vendors even shooed them away with a kick as the turkeys clucked and tried to cadge crumbs.

In the Costa Rican rain forests, I found botanical jokes. Impatiens blooms extravagantly there. Not the neat impatiens of our North American herbaceous borders, but two-, even three-foot-high muscular impatiens. It grows throughout Costa Rica, blooming its heart out, an exotic invader from the Far East.

The rain forest is full of house plants. All our homely philodendrons, begonias, dieffenbachias, bromeliads, and ficuses are rain-forest plants, and in their natural habitat they grow to extravagant heights, are lush in leaf and sturdy in stem, radiating good botanical health; they show you what they can do when they are growing in situations that suit them.

Of course most of the plants are unfamiliar, with leaves that, to a North American, seem peculiar in shape, too big, too weird. Big trees brace themselves with buttress roots; other plants turn epiphyte and compete for places on the tree limbs. The epiphytes have epiphytes themselves. Strangler figs surround, choke, twine upward toward the light, throw roots back down to the ground, and start vining upward again. Except for the occasional flowering orchid or bromeliad, and, of course, those impatiens, the rain forest is a monotone of green — layered, intertwined, inexorable masses of greens, subtle and bold, every size of green, every shape, form, and hue of it. It is the Goldberg Variations of green.

I was so overwhelmed with green strangeness when I was in

Tikal that it was with relief and a sense of escape one day that I climbed the stairway and highest ladder of the highest of the high temples at the Mayan ruins and found myself above the rain-forest canopy, in the sunshine, out of the dim green dusk of the world below, which was puzzling me entirely too much. Up there, 145 feet above the ground, on top of what was, until North American skyscrapers were built, the tallest man-made structure in the Americas, I could look down on the treetops. I saw monkeys, howler and spider, zestfully and confidently crisscrossing the forest in troops. On the ground they had peered at me intently. Up in their home territory, they ignored me. I could see for miles and miles across Guatemala, and it all appeared to be green treetops. I could see no break anywhere.

I had taken that trip seven years before I went to Belize, and although I'd seen a number of rain forests in the interim, I hadn't made it up into the canopy again, though I had wanted to.

At the Blue Creek preserve and field station there is a walkway into the canopy, and I was given directions to it. They were clear enough. Take the walkway at the cabbage bark tree to the provision tree. There I would have three choices: turn right and walk to the Santa Maria tree, or turn left and walk to the bullywood tree, or climb straight up to the top of the provision tree. Clear, yes, but for starters, the walkway between the first two trees was 60 feet in the air, spanning the creek rushing below it, and the other pathways were higher still. They were all hanging bridges. No easy, hospitable, hand-hewn Mayan stairways here.

I stood at the base of the cabbage bark tree, which leaned out over the creek, and looked up the series of aluminum ladders and the wooden cross-struts bolted to the tree. The ladders disappeared into the leaves of the rain forest above, but I knew they led to the first hanging walkway, my only passage to the other walkways and the platforms 80 and 125 feet up in the canopy on the other side of the creek.

Feliciano Bol, chairman of Blue Creek Village, allows only two people at a time onto the swaying bridges, and he was going up with me. I asked him, as he strapped me into a safety harness and handed me a hard hat, if he would take his grandmother up there. He laughed at the thought. "No," he said. "She'd be frightened."

"Well, let's pretend I'm your grandmother this morning," I proposed. I could feel the adrenaline pumping, but I'd wanted to walk into that alien world for seven years. This particular rain forest is the year-round home of tropical birds and the winter resort for many North American ones. There are howler monkeys here, too, and lizards, kinkajous, and bats. But above all, the forest is home to thousands — millions — of insects, spiders, mites, and a host of other invertebrates, most of them unknown to science.

It is only in the last couple of decades, as techniques have been developed to get us up into it, that canopy science has become a field of study. Those techniques include gigantic cranes, sledges that float from hot-air balloons, baskets hauled up after a rope has been shot over a high limb, and walkways such as the one at Blue Creek. None of the methods has proved entirely satisfactory — we are not arboreal, as my nerves proved — but they have given us the first tantalizing glimpse into this little-explored world.

Feliciano explained how the clips on my safety rope and strap worked, cautioned me to have one in place, always, before I released the other, and instructed me to push ahead the Prussik knot, a looped noose fastened to the rope that ran up the side of the ladder. I clipped the strap from my harness to the rope so it would break my fall if I slipped.

The bolts on the ladders looked secure, the safety devices sensible. I have climbed trees all my life, and I was Feliciano's honorary grandmother. I started up behind him, shoving my Prussik knot ahead. At the top I hooked my safety rope onto the overhead bridge cable, unhooked the strap, told myself not to look down ("Remember," an experienced rock climber at the field station had told me, "nothing matters after the first thirty feet"), and eased out onto the swaying bridge of metal rungs linked by cables.

A few steps out, I realized it was fun. Fear faded and curiosity took over. It was early morning, the sun was shining in a blue sky, beginning to dry out the night's rain from the bridge surface. Bats, which I'd watched vacuuming the air of mosquitos at five A.M., were roosting in trees and caves, but I could hear a nightingale wren singing ahead in the provision tree. Mists were rising in cloudlets all around.

On the other side of Blue Creek the bridge connected with the other hanging walkways. It required a bit of tricky footwork to get to the slatted wooden walkway to the right, but I followed Feliciano's directions slavishly and went out onto it. He followed behind, and our paired walking made the bridge undulate. In the Santa Maria tree there is a platform at eighty feet; sitting on it, I allowed myself to look down for the first time. The sun sparkled on the water boiling over the low waterfall upcreek.

The day before, I'd hiked with a student group from a Seattle prep school that was staying at the field station cabins. We'd climbed to the source of the creek in a very big cave at the head of a gorge in the merest hint of the foothills of the Maya Mountains. That evening I was a guest in the village home of Delfina and Sylvano Sho. Sylvano works at the field station as the cave expert and guide. Delfina and Sylvano's house is made of vertical wooden slabs and is roofed with thick thatch. The evening meal had been cleared by the time I got there, but the aroma of good food cooked over a fire lingered. Sylvano gave me the household chair to sit upon. As the fire died down, the one-room house was lighted by a candle — electric power does not reach Blue Creek Village. I asked Sylvano to tell me about the cave I had seen that day.

Fifty-two years earlier, Sylvano's father had moved to the banks of the creek to farm. Exploring upcreek one day, he had come to its source in the cave, where he found a group of unfamiliar Indians living. They spoke a language he couldn't understand and threw stones at him. Over the next few years Sylvano's father and uncles tried again and again to visit the people but were always driven away by a hail of rocks. They did, however, establish that the group had only stone tools and that they had musical instruments, drums and flutes, which they used in ceremonies conducted by a man Sylvano called a priest. Caves were regarded as sacred to the ancient Maya, so it made sense to Sylvano that they would conduct ceremonies in one. One day the people disappeared, Sylvano's father told him. Sylvano thought they had left because they did not want to be observed.

When he was fourteen, he began to explore the cave and found that it was part of a system that branched throughout the mountain and came out on the other side. In one place he found human bones and wondered if they might have been the remains of the band of people his father and uncles had seen. He also discovered artifacts, statues, something resembling a stone thermos bottle, bowls, and tablets from ancient Mayan days. Writing on the tablets, deciphered by an archaeologist, gave the name of the cave, Hokeb Ha, meaning Source of the Beautiful Blue Water. Sylvano said that when he was a youngster, Blue Creek was much deeper and he had once seen a crocodile in it, but farming nearby had made it silt up and become shallower.

As I sat on the leafy platform, I saw no crocodiles in the creek below, but I watched iguanas struggle out of it. Iguanas seem only marginally more suited to the treetops than we are and, lacking safety harnesses, they often tumble down from branches into the water and have to scramble for the bank.

At eighty feet, this platform looked down on the lower-story trees as well as up at more canopy above. A combination of ladder and climbing bolts led up to the 125-foot platform. There we were right inside the canopy and could see into the complexity of growth — trees of many species, epiphytes growing upon them, orchids and bromeliads, some in scarlet bloom, and strong lianas, draped vines that became bridges for small animals from one tree to another. Strangler figs had enveloped some trees so completely that the tree itself could no longer be seen. I heard a brown jay scolding and saw a toucan playing hide-and-seek with us over at the bullywood tree.

Feliciano is the local bird expert. Not only does he know all the natives and their songs, but he can identify the North American migrants, too. He didn't learn from books, for he didn't have any; he just watched the birds themselves and asked questions of North American birders who visited the field station.

Insects filled the air around us with their humming. On the ground I hadn't found a peripatus, but I'd seen fireflies. I'd watched leafcutter ants climb the trees out of sight and return with bits of leaf. They carry the leaves to their nests, where they grow fungus on them for food. I'd seen tall, mounded termite nests and their constructed tunnels clinging to trees, protecting the termites from drying air as they browse upward. Some of the tunnels extend high into the trees. In the lodge dining room, I'd watched a tarantula strolling possessively across the floor (it was in its own place, after all), and I'd heard that it was wise to check one's shoes for scorpions before putting them on.

The best available estimates say that 80 percent of tropical insects remain unknown and undescribed, so most of those that buzzed and crept around me probably have never been identified; their lives are not known to us. But there aren't enough scientists to collect them and even fewer taxonomists to keep up with the samples that *have* been collected and sent back to research institutions.

A few reports on invertebrates living in rain-forest canopies in widely scattered places have been published — but Blue Creek is not among them. A very preliminary insect count from the Blue Creek treetops has been made by Nathaniel Erwin, director of the Smithsonian's Insect Zoo. He is so busy with administrative duties that he has not yet had time to tabulate his census, let alone identify what it contains. All he could tell me was that he had found a lot of moths, particularly miner moths, as well as katydids, beetles, and a whole bunch of ants. "One thing was particularly puzzling about the ants," he said. "There was one group of species in the trees during the daytime and an entirely different group at night." (Nate is a personable, earnest young man who completely won my heart one day at a demonstration of bug cookery, when he tried to interest a timid spectator in a dessert he had made by asking, "How can you miss with chocolate and crickets all rolled up into one?")

Some studies of Blue Creek's plants, reptiles, and a few mammals have been done, but Nate is the only one who has studied its insects. And no one has taken even a cursory look at any of the other invertebrates. Sam James had told me that some species of earthworms live out their entire lives in bromeliads growing along the limbs of rain-forest trees, so they might be found at Blue Creek, as would mites. A study done in a rain forest in Australia found that half a million mites lived on the leaves of an average forty-five-foot-tall rosewood tree.

Nate had found a lot of beetles, and on the platform I watched for them, too, but found them beyond my modest identification skills. Beetles, in a sense, are the beginning of canopy science. Terry Erwin, who is not related to Nate, works for the Smithsonian as the resident research specialist on beetles. Some twenty years ago he began his tropical forest canopy fogging studies, spraying an insecticidal mist upward, then collecting and identifying what fell into his catch nets below. After discovering a great many new species in his samples, he estimated that in addition to the 370,000 known species of beetles, there may be 10 million more awaiting discovery, most of them in the rain-forest canopy. Although his first estimate has been disputed — it may be too high, it may be too low — it brought the world of the forest canopy to the attention of scientists and opened it up to study.

People who deal with vertebrates often assume that invertebrate zoologists are just picky and overzealous when they talk about the difficulties of sorting out who is what, and a little mad when they talk about the large numbers of species. Vertebrates are big animals with big ranges. And with birds, for example, it is fairly easy to tell one from another: they look different. But among invertebrates this is often not the case. They are small animals, often have extremely small and restricted ranges, and one species often looks much like others, even though they are very different biologically, live differently, and do not interbreed.

Those of us from simpler places find the rain forest puzzling; it may be that we have an intuitive sense of its diversity without fully understanding it. Even scientists are awed by the wealth of species. The mangrove island off the coast of Belize where I spent a day with Candy Feller is supposed to be depauperate, meaning, in biological jargon, that it has few species. The brackish water and the unrelenting exposure to the elements are supposed to make mangrove swamps difficult places for most plants and animals to make a living. But Candy is challenging that notion. She showed me the telltale tracings of leaf miners in the leaves of a red mangrove tree. The larvae of miner moths had been tunneling through the leaves, eating their insides. She has discovered at least four other species of miner moth in addition to the one whose traceries we saw. They all belong to the same genus but live in other parts of the mangroves, such as the aerial roots or new growth tips or twigs, and are biologically distinct one from the other.

As I sat on the high platform I wondered, if one can find five species of a single genus of moth miners in a stressful place like a mangrove island, how many species might there be in a lush and diverse place like the rain-forest canopy? A preliminary botanical survey of Blue Creek, written up by Margaret Lowman, a botanist from the Selby Botanical Gardens in Sarasota, Florida, found that on average, each transect of a little under twenty square feet had 127 different species of plants growing on it. I could see leaf-miner tunnels in the leaves of many different plants; it would not be surprising to find that each species of plant had its own species of leaf miner, or even several of them.

Whatever the numbers, everyone acknowledges that rain-forest canopies allow for a much greater number of microhabitats than a mangrove forest does and that they are just the kinds of places where many species of small animals, particularly invertebrates, become established and thrive. Coral reefs are another place where enormous numbers of animal species make their homes, and the deep sea may also be surprisingly rich in unknown creatures. The long-accepted notion was that the deep ocean floor was too poor in food to support much life, that most marine life was to be found in the water's upper regions; this led to an estimate of some 160,000 marine species in total. But improvements in sampling machines have shown that there are many shifting microhabitats in the deep oceans and that the snowfall of debris from above is actually a very rich source of food. Lots more kinds of life live down there than we ever dreamed of. J. Frederick Grassle, of Woods Hole, who has sampled the deep cold waters of the northwest Atlantic, off the coast of the United States, estimates there may be one million, or even ten million, species of animals living in the oceans altogether.

Like Terry Erwin's figures, Grassle's have been disputed, but no one argues the point that a good deal more is going on in the world than we ever knew; one authority, E. N. K. Clarkson, the University of Edinburgh paleontologist, is of the opinion that there are twice as many species living today as there were in Paleozoic times.

At my moment in time, on the platform in the rain-forest canopy at Blue Creek, butterflies flittered in the sunlight as Feliciano told me about the brilliant white morpho, a sun-loving butterfly he had recently spotted in the canopy's bright openings. I recognized some of the bees humming about us as the tiny stingless kind that the Mayans kept for honey and wax. I asked Feliciano about them. He said that several families in the village keep them and that their wax has an important use. When someone has taken a bad fright, he explained, beeswax can cure it. Feliciano himself was cured when, as a child, he slipped from a rock into the creek and became terrified of water. The bush doctor came and drew crosses in melted beeswax on his forehead, arms, and belly. When the wax dried and the bush doctor scraped it off, his fear lifted with it.

Years ago I read a Mayan story from the Guatemalan highlands about the creation of those stingless bees. It was said that an old man had two beautiful daughters. The story doesn't tell the fate of the first, but the other caught the sun in the shape of a humming-bird and fell in love with him. Each night, after the sun's duties were over, the hummingbird would come to her. The father grew suspicious and questioned her harshly. She lied and told him she had no nighttime visitor, but he did not believe her and began to lie awake at night to watch. Before many nights passed, he discovered the two together. Furious, he called upon his friend the thunder to kill his deceitful daughter. Friend thunder obeyed him. The hummingbird, in his grief, gathered up his lover's bones and put them into a pottery jar, which he gave to an old woman to keep. After a while the old woman heard a noise in the jar, which so frightened her that she covered the jar and hid it in a corner. The hummingbird returned and asked for the jar, but the woman would not tell him where it was. The grieving hummingbird searched and found it. He could hear the noise, too, but it did not frighten him. When he looked inside he found that his lover's bones had turned into bees, which would give honey and wax to anyone who would take care of them.

Feliciano and I had been more than an hour in the treetops, and I knew that some of the Seattle students wanted to climb up, too, so we left our high perch. The sun — I saw no hummingbirds —

had dried the walkways and my descent was on surer feet. Feliciano positively ran across the hanging bridge; he was hungry and was headed for breakfast. I felt exhilarated as I unstrapped myself at the base of the cabbage bark tree and followed him into the lodge, where breakfast waited for me, too.

I spent several more days in Blue Creek, damp rain-forest days. It rains, on the average, 170 inches a year there. On my last morning, as I stood in the village at the creek's edge, waiting for the pickup-truck driver who would take me over thirty miles of bad road to the landing strip at Punta Gorda, I realized that I smelled mightily of mildew. But I was entirely happy. I'd walked in the rain-forest treetops with beetles and bees and birds and bats.

READINGS

Grassle, J. Frederick. "Deep-Sea Species Richness." *American Naturalist* 139, no. 2 (1992).

Lowman, Margaret D., and Nalini M. Nadkarni. *Forest Canopies.* New York: Academic Press, 1995.

May, Robert, "How Many Species Inhabit the Earth?" *Scientific American,* October 1992.

May, Robert, John H. Lawton, and Nigel E. Stock. "Assessing Extinction Rates." In *Extinction Rates,* ed. John Lawton and Robert May. New York: Oxford University Press, 1995.

Thompson, J. Eric. *Maya History and Religion.* Norman: University of Oklahoma Press, 1970.

Tropical Ecology Workshops Manual: Belize. Sherborn, Mass.: International Zoological Expeditions, n.d.

Wilk, Richard, and Mac Chapin. "Belize: Land Tenure and Ethnicity." *Cultural Survival Quarterly* 13, no. 3 (1989).

10

We must have . . . objects themselves to serve as the factual basis for knowledge, the final arbiter in matters of contested identity or meaning, the "ground truth" that underlies our understanding of the world we inhabit.

— Anna K. Behrensmeyer, acting associate director
for science, National Museum of Natural History

DURING A STRETCH of Washington months I got to know, tolerably well, a big beautiful spider, blotchily orange and tan with darkly banded legs. Each day she spun a fine new round web somewhere in the garage or, occasionally, on the back porch. She usually sat off the web, hidden against the ceiling or a protecting beam. When moths and flies blundered into her trap, though her eyesight was none too good, she could feel the vibration of one of the guying threads and would rush out onto the web. She would eat the first of her catch and wrap the rest in silken winding sheets to keep for later. I always tried to avoid tearing her web and save her repair work, but I needn't have bothered. Jonathan Coddington, curator of spiders at the Smithsonian's Natural History Museum, told me that she ate each day's web and reprocessed the protein in it. Within twenty minutes of munching it down, she was capable of recycling the silk. Another individual of her same kind came to stay at the illustrator's house and can be seen below.

Her common name is barn spider, her scientific one *Araneus cavaticus.* Her genus, *Araneus,* includes more species than any other spider genus. It, and the name of her family, Araneidae (the spiders who spin orb-shaped webs), as well as that of her order, Araneae (all spiders), and even her class, Arachnida (spiders plus a lot of their kinfolk: ticks, daddy longlegs, scorpions, and such-like), echo the name of Arachne, a Lydian princess. Arachne was such a skilled weaver that the goddess Athene, who fancied her-self the best at that art, grew jealous of her. The goddess, deter-mined to find technical fault with Arachne's work, examined one of her tapestries, into which Arachne had provocatively woven a

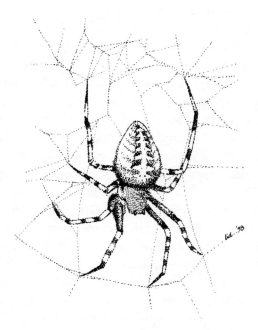

A barn spider, *Araneus cavaticus* (2 × lifesize).

story of the love affairs of the gods. But Athene could find no technical fault, none whatsoever, no matter how hard she tried. That made her even crosser, and in a fury she tore up the work. (No one ever said the gods were nice.) The princess, terrified at having aroused the fury of a goddess, hanged herself with a rope from a rafter. Athene, still irritated and not sufficiently satisfied with the mere death of an irreverent rival, turned Arachne into the animal she hated most of all, the spider, and transformed her rope into a cobweb.

As spiders go, *A. cavaticus* is famous. She is the heroine of E. B. White's book *Charlotte's Web,* in which she saves the life of Wilbur the pig by writing messages in her web. I never did see TERRIFIC embroidered in the garage, but for someone like Jon Coddington, a specialist in spider behavior and taxonomy, the barn spider's web does have messages.

Today's taxonomists aren't just in the business of assigning names to newly discovered animals; they are also folding into the classification scheme everything heritable that is known about a particular animal and its relationship to all other animals. They ponder zoological principles and ask questions. What constitutes spiderishness (order Araneae)? How does one family group of spiders differ from or resemble the others (families Araneidae and Theridiidae)? What is the evolutionary relationship between the groups of spiders that spin different kinds of webs (Charlotte and all the rest)?

In asking and answering these questions, Jon settled a long controversy about spider phylogeny, or evolutionary trend. Spiders spin a variety of webs, each peculiar to its kind. Sheetweb spiders (family Linyphiidae), for instance, make the webs that look like little flattened hammocks that you often see in the shrubbery in the early morning, frosted with dew. Spiders of the family Theridiidae are the ones who spin those cobwebs that people tidier than I am dust out of basement corners. My *A. cavaticus* and others in the family Araneidae spin the familiar orb-shaped webs of Halloween art.

It had long been assumed that orb webs were the highest achievement of spiderly craft, the most advanced. Such webs look so perfect to our eyes that they were thought to be maximally efficient, possessing such adaptive value that their making had evolved independently in different groups of spiders. They were certainly worthy of a spider who could lay out SOME PIG in her web.

After careful observation, however, Jon concluded that rather than being the ultimate web, the orb web is primitive, ancestral, the kind of web the first spiders spun. The scorned, messy-looking (to our eyes) cobweb is more elaborately engineered, denser, safer for the spider, and more efficient for trapping prey, which is, after all, the purpose of a web. All spiders are carnivores. So the cobweb actually evolved from the orb web, as did other, more specialized webs. The earlier spider classifications that separated orb weavers into different groups based on anatomical differences have been generally scrapped now, Jon told me, because putting all the orb weavers together better represents what we know about them.

I sat in Jon's office on the third floor of the Smithsonian's Natural History Museum as he explained all this to me. The sun flooded through windows overlooking the Mall beyond. Jon, a tall, lanky, rumple-haired young man with glasses and a lopsided smile, wore an open-necked shirt, khakis, and Birkenstocks.

The Natural History Museum is one of the most popular stops for tourists visiting Washington. More than five million visitors come there every year. Afterward they probably remember the elephant in the rotunda, the nice lady at the information desk, or dinnertime for the tarantula at the Insect Zoo, where Nate Erwin presides. What they will not remember, for they will not have seen it, is the undisplayed 99 percent of the collections of 122 million "objects" like those millipede specimens Ron Faycik helped me with or the earthworm specimens out at the museum's support center in Maryland, where Sam James worked. Nor do they see the four hundred scientists and technicians, people like Klaus Ruetzler or Jon, who work with those "objects." Together, the scientists, staff, and collections make up the largest natural history museum in the nation and one of the preeminent research institutions in the world. Security in this nonpublic part of the Smithsonian is tight, as it is in all government buildings, and casual visitors, for obvious reasons, can't just wander in and start pulling drawers out of collection cabinets, as Ron did that day with the millipedes. You need a special invitation, and even then you have to go through security, turn over your driver's license as bond, assume a visitor's badge, and have an escort. The escort is needed to find your way through the mazelike hallways threading between collection cabinets and tiny, crowded offices. The entire behind-the-display area smells faintly of moth crystals and preserving fluid. Ever since I'd seen those drawers filled with millipedes I'd wanted to come back and learn more about how the collections were put together. Jon, the first spider expert to be appointed at the Smithsonian, is giving the spiders a good and much-needed sorting out.

Zoologists' offices are always stuffed with books, files filled with papers, and stacks of journals, and Jon's is no exception. But it also has fanciful wire sculptures of spiders in their webs and real, but dead, spiders floating in alcohol in collection vials. Spiders, spiders, spiders. Everywhere. He showed me a flat container with a web in it but without its spinner. The web had been made by a brown recluse spider, whose bite is as dreaded by many people as that of the black widow. He had been working with researchers from Du Pont who were interested in developing threads with the strength and other qualities of spiders' silk.

The researchers had been concentrating on the golden silk spider, a neotropical orb weaver famous for its strong web. "But I

think there are others worth studying," said Jon. "The brown recluse may be a better bet." The web in the flat container looked
dense. He showed me some scanning electron microscope photos
of the unusual, blunt spinneret that is part of this spider's silk-
making equipment. The silk comes out flat and ribbon-shaped,
not threadlike, as it does with other spiders. "We don't know
much about its qualities yet, but it is worth investigating," Jon
said.

Jon's desk was covered with collection vials containing specimens he had found in East Africa. He spends about half of his time
out in the field collecting and studying spider behavior. He usually
identifies them only to the family or genus level. "I leave the deter-
mination of species to someone who specializes in that group," he
told me. He notes down the collection data and then passes the spiders along to a colleague, who enters the information into the
computer record and prints out labels that identify them and tell
the when and where of their collection. Groups of vials containing
the new specimens are packed into straight-sided, alcohol-filled
half-liter bottles closed with rubber gaskets and metal clamps like
old-fashioned canning jars.

Jon took me back to a room filled with tan metal cabinets that
look like map cases with fat drawers. The millipedes had been
stored here before being transferred to Maryland, but now the
drawers were filled with the half-liter jars of spiders, arranged
alphabetically by family and subarranged by genus and species, if
known, and geographic origin. Jon pulled out a drawer containing
A. cavaticus specimens. There were two vial-filled jars of them, but
many more of other species of *Araneus*. All told, Jon estimated, the
museum's spider collection contains some 116,000 specimens.

The entire entomological collection, which includes not only
spiders but insects and myriapods, contains thirty times that number, or 31 million specimens. "It is," Jon told me, "one of the most
inclusive and accessible entomological collections in the world.

The only other one of equal importance is the one at the British Museum."

Specimens, preserved and classified, make up what is called a synoptic collection. That means it does not include a series of every single known spider, but is rather a synopsis of them, with representatives of species or at least of the higher taxon — genus or family — that includes that species. "Our museum," said Jon, looking around it, "contains the greatest synoptic collection of spiders and other terrestrial arthropods on earth. This is where biodiversity can be studied." Each year some nine thousand scholars visit the museum to make use of collections such as this.

One famous researcher who visited was Hirohito, better known as a divine emperor than as a marine zoologist. When he came to Washington on a state visit in 1975, he was keen to examine the museum's coelenterates, a sampling of the hydra, jellyfish, sea anemones, and corals of the world, to clarify some of the species in his own collection. So keen was he, in fact, that he overstayed the brief time his State Department handlers had allotted and, to their distinct annoyance, refused to leave until he was done with his work. Ellis Yochelson, the museum historian who tells this story, commented, "This may be one of the few times that protocol has given way to natural history."

Researchers can borrow from the museum. In 1996 more than 140,000 specimens from the entomological collections were sent out on loan. "Let's say someone is trying to revise a group," Jon said. "He has two females at a certain taxonomic level and no males. He needs a series to make the identification, so he borrows what we have." A loan is normally made for two years, but it can be renewed, and sometimes specimens are kept ten, even fifteen, years by qualified researchers.

Back in his office, Jon pointed to the fancywork of the brown recluse and said, "Good taxonomy has predictive value, which is often useful. For instance, there is a spider in South Africa known as *Sicarius* [the name means "murderer"] that can give a serious bite. No one has done much work on it and we don't know about the species, but we think the genus *Sicarius* is sister to genus *Loxosceles,* to which the brown recluse belongs. We *do* know rather a lot about the brown recluse, so we can make predictions about the biology of *Sicarius* and know how to treat the bites."

Bad taxonomy, on the other hand, can be expensive. Many millions of dollars have been spent in the attempt to eradicate the gypsy moth in this country; it was imported from Europe a little over a century ago as a consequence of what we now know was mistaken taxonomy. Leopold Trouvelot, a French amateur naturalist and astronomer, was using the gypsy moth in an experiment to develop a better silkworm when his European specimens escaped from his laboratory near Boston. Undeterred by natural predators, gypsy moths have been eating their way through our eastern forests ever since. In Trouvelot's time the gypsy moth was classified in the genus *Bombyx,* that of the silkworm, which was, and still is, *Bombyx mori.* The gypsy moth was, but no longer is, *Bombyx dispar* (which means "silkworm with males and females that look different"). Trouvelot thought that the two moths were close relatives because of the taxonomy and may even have considered attempting to interbreed them. He probably would not have experimented with the gypsy moth (kept in fair check in Europe by its natural predators) had he known it by today's name, *Lymantria dispar.* The genus name means "destroyer." Today's taxonomy takes into account its evolutionary history and puts it, along with the other tussock moths, in the family Lymantriidae, a considerable relational distance from the family of silkworms, the Bombycidae.

All the invertebrate zoologists I have ever talked to have lamented the state of taxonomy in his or her own particular field. In many the basic relationships have never been worked out at all, and in others they need to be brought up to date — the process is called revision — to reflect evolutionary thinking and to incorporate the considerable new information that tools such as molecular analysis are able to give. One eminent authority, E. N. K. Clarkson, has estimated that of all the described species, only about one-sixth are "good," that is, properly defined in the sense that we know for sure that its members don't interbreed with those of other species.

"The folks who think taxa are hypothetical or human constructs," Jon said, "are just wrong. Many taxa on earth are programmed to need, hunt, and depend on other particular taxa. So flies, molds, viruses, worms, bees, and so on 'agree,' as it were, with human taxonomists. When specific identity is important, it usually turns out that we are all (humans and whatever else) in agreement. We are *discovering* a natural order that exists independently of the observer."

But in addition, the bits and pieces of life are so numerous that we need to order and classify them before we can think about them. Our sort of brain cannot handle the world in the raw. We have to arrange all the bits into piles, and if there are too many piles we arrange those into clusters. Without ordering systems, which is what taxonomies are, we can't think, live, or work with our world. We would find it hard to make our way through a shopping list at the grocery store if we didn't have mental categories. Without, say, the category "orange," we would have to remember each time we shopped that we wanted those orange-colored things with sweet pulp inside and not the yellow-colored things with tart pulp inside. What is more, we want the oranges and lemons grouped together so we can think of them as "citrus." And please put all the fruits and vegetables together in one big section and call it produce, rather than mixing them with the coffee and kitty litter. That's taxonomy. And the categories, whether they are called oranges, citrus, and produce, or species, genus, family, order, and class, are called taxa. The singular is taxon.

From Aristotle to Foucault, the world's heavy (and sometimes not so heavy) thinkers have seized upon taxonomy and ordering systems as an intellectual and practical joy. It is so much fun to create neat systems! In another part of my life I was a librarian. Librarians make a profession out of arranging things. They put those mysterious letters and numbers on the spines of books so that they can be arranged on shelves in order of the subject of their contents. The usual systems of arrangement in this country are the one used by the Library of Congress, a couple of letters and then a number, and the Dewey decimal system, numbers only. But other systems have been invented by classificatory minds at work. I remember learning in library school about one particularly stunning system called Faceted Classification, invented by an Indian whose name I no longer remember. In his system, a book was assigned a number that put it in relationship by subject not only to the books on either side of it on the shelf but also to those on the shelves above and below. The only trouble with his system, perhaps considered a minor one to such a systematist, was that books could never be removed from the shelves because that would break the pattern of relationship. In hindsight, many historical attempts to arrange the natural world seem as contrived and bizarre as the Faceted Classification System, but then, two hundred years from now our systems, reflecting as they do the way we look at the world, may seem equally amusing.

Aristotle's approach to animal classification, using a single characteristic (those with blood and those without, those with hair and those that are hairless), dominated scientific thinking down through the centuries and still, in many ways, influences the way we look at the world to this day. Linnaeus was also a single-characteristic classifier. He divided insects up by their wing style, for instance. And we still use several of his names for the insect orders, such as Lepidoptera, the scaly-winged, for butterflies and moths. When you use just a single characteristic you can arrange things by other single characteristics, too. Linnaeus thought wings were important, but a pupil of his, Johann Christian Fabricius, thought that when it came to insects, jaws were all. He regrouped the insects into piles by the kind of jaws they had (in the Linnaean world it was believed that there were only nine thousand species of all living things, so it was comparatively easy to restack the piles). A relict of Fabricius's classification scheme lives on in our name for the order to which dragonflies belong, Odonata, the tooth-jawed.

We've mostly discarded Linnaeus's original classification, but we've kept his most important invention, which was the custom of giving everything a two-part scientific name. Those names may seem hard to understand these days, when few people are familiar with Latin, but they are an enormous improvement over the kinds of names given before Linnaeus. Before his time, for example, the honeybee was known as *Apis pubescens thorace subgrieseo abdomine fusco pedibus posticis glabris utrinque margine ciliatis.* But clever Linnaeus saw it could all be done with just two words. The second one, the species name, isolated animals that were like no others; the first word, capitalized, the genus name, grouped individual species with others they were similar to. So the honeybee became simply *Apis mellifera,* the bee of sweets.

Other thinkers had other ways of looking at the world. The eighteenth-century French naturalist Michel Adanson, believing that single-characteristic ordering systems were "artificial," developed a Universal System, which made use of all known characters. He believed that such a system was more natural, but in his time it was considered too unwieldy and Adanson was labeled an eccentric. He sounds like an endearing man. Toward the end of his life he asked that his grave be marked with a garland of flowers from the fifty-eight plant families he had elucidated.

Many others looked for "natural" systems, too. They believed that life's mystical unity would be revealed if only the presumed

The family of crows and ravens according
to the Quinarian system.

affinities between the bits of it could be arranged in the correctly discovered configuration. One such system was called the Great Chain of Being, a ladderlike arrangement, with fire, water, and air at the bottom, extending through the "lowly" plants and animals to man, triumphant on the top step. Then there were the Quinarians, who held that all life could be divided into systems of fives. They arranged five circles in a greater circle and found affinities where the circles touched, or "osculated." Each circle could be subdivided into additional fives, such as the one shown above, a circle of five-pointed stars of the family of crows and ravens. Symmetry is maintained by declaring that certain unfilled points on some of the stars represent *undiscovered* crows (which tells more about the human mind than it does about crows).

Goethe is one of the best known of the thinkers who hoped to work out a natural system for ordering the world. He coined the word "morphology" to mean the study of organic form. His scientific interests were primarily botanical, and, with a bow toward Plato, he believed that all plant forms were shadows of an Ideal, a dawn-plant, an *Urpflanze,* and that a classification scheme could elucidate what that Ideal might be.

Goethe notwithstanding, no one before Darwin considered organisms except as they existed, fixed in form, at the particular moment. The most empirical classifiers examined body parts, measured genitalia, compared color, speculated on the function of organs of dead specimens. Darwin's evolutionary thinking forever changed ordering schemes by introducing the principle of time, of history, into natural history. An animal's origins and lineage began to be taken into account.

Today's taxonomy goes by the name of cladistics, and the pictorial representation of it is called a cladogram. The name comes from a Greek word, *klados,* for a tree's branching. It is not simply a pictorial classification scheme, however; it is more like a densely packed encyclopedia of current biological knowledge, subject to revision and editing as new discoveries are made and readable by anyone with some understanding of the animals diagrammed. Cladograms are family trees showing which animals are "sister groups" or have degrees of cousinship; who has kept certain anatomical features or ways of doing things; who invented new ones. The cladogram on page 165 shows the relationships within an infraorder, the Araneomorphae; the barn spider's family, the Araneidae, can be spotted eleventh from the bottom.

Cladistics is not so much a fixed classification as it is a process of analysis by which taxonomists try to discover an organism's evolutionary tree, based on traits shared with other organisms. A given taxon is defined by the unique heritable traits, thought to be of recent origin, that set its members apart from all other taxa on the

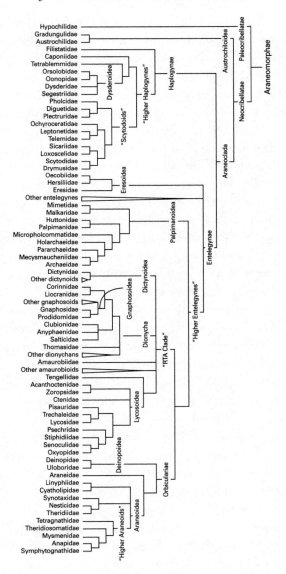

A cladogram showing the family tree of the spiders.
Adapted from Jonathan Coddington, "Systematics and Evolution
of Spiders," *Annual Review of Ecological Systematics* 22 (1991)

family tree. The unique characteristics of a taxon evolved at different times over its evolution, so the charting of the appearance of these traits results in a branching diagram on which ancestral species are inferred at the branches.

The analysis makes use of everything heritable that is known about the animals within the group (the ghost of Adanson walks, this time equipped with a computer): bodily form, biochemistry, manner of living, molecular profile, development. Jon pulled out from a pile of papers on his desk a computer printout that he called, proudly, "*the* matrix." It contains 49,000 "potential cells," bits of information that he and his colleagues have gathered on spiders, which serve as the characters for cladistic analysis. They are coded so that they can be sorted and compared by computer.

At any taxonomic level some characteristics are considered primitive because they are shared with other, somewhat similar animals; common to all, they are therefore "original" and ancestral to all. So for that particular taxonomic division, they are ignored. For instance, all arachnids have eight legs, so if you are classifying spiders the character "having eight legs" would be ignored, for ticks and daddy longlegs have eight legs too. But other characteristics, peculiar to spiders alone, such as "having spinnerets at the nether end" or "having poison fangs, a head and then the rest, attached by a narrow waist," *would* be considered. Those help define spiderishness. But *within* the order of spiders, narrow-waistedness and spinnerets-at-the-end would be ignored, considered primitive, because all spiders have them. Other, more special characteristics would be considered derived, unique.

I asked Jon what would separate the more efficient and more lately evolved cobweb spinners, the Theridiidae, from primitive orb weavers like my *A. cavaticus*. He said that cobweb makers hurl big droplets of sticky silk to attack their prey, while *A. cavaticus* can only hang sticky silk in its web. The hurling of sticky silk is shared by all cobweb makers, whatever their species. It is unique, a derived character.

Cladistic analysis is not even half a century old. Its basic under-
pinnings, called phylogenetic systematics, were worked out by a
German entomologist named Willi Hennig, who died in 1976.
Hennig knew that change, flux, is characteristic of life, that an insect may be a little, squirmy, wormlike thing when it is young and
in larval form but a winged beauty as a mature, reproductive adult.
He knew that many marine invertebrates, such as the barnacle or
sponge or coral, float freely in the plankton when immature but
change in form and settle down as firmly as a plant after they grow
up. He also knew that classification systems in the past had been
based on dead specimens, fixed for all time in preserving con-
coctions at a particular stage of life and that, as a result, sometimes
the same animal, larva and adult, had been given separate species
names. He wanted to develop, not so much a classification, but
what he called a "general reference scheme of biology" that took
into account change, considered entire lives and the way they
were lived. He wanted to incorporate paleontology, geology, and
biogeography into biology.

Hennig was held prisoner by the British in Italy during World
War II, and it was in prison that he started writing down his
thoughts on the subject in a notebook. His massive *Grundzüge
einer Theorie des phylogenetischen Systematik* was not published,
however, until 1950 by an East German publishing house (he lived
in West Berlin but worked in East Berlin) on "poor paper . . .
written in a style of language which is difficult to read, even for
Germans," according to Dr. Michael Schmitt of the Museum
Koenig, who is working on a biography of Hennig. Dr. Schmitt,
with whom I corresponded, also wrote that Hennig's work was
"overloaded with philosophical . . . stuff." His ideas were pretty
much ignored after the book's publication. In 1961, as a personal
protest against the building of the Berlin Wall, Hennig quit his
job with the German Entomological Institute and subsequently
became director of phylogenetic research at the State Museum of
Natural History in Stuttgart. A decade later, an English-language

version of his work appeared as *Phylogenetic Systematics,* and scientists began to use his system. Today, in revised form, it is the standard tool for analyzing relationships in the living world, used by museum and university scientists nearly everywhere. Hennig did, indeed, create a "general reference scheme of biology."

I asked Jon what he thought about the huge numbers of species estimated to exist and how those estimates affected a museum curator. "Actually I'm not too interested in species," he said with a dismissive wave of his hand. "It's easy enough to extrapolate those numbers by keeping track of how many new ones turn up with each family or genus actually collected on successive expeditions to any one place. There are something better than 36,000 known species of spiders and perhaps another 100,000 as yet unknown. But the higher taxa represent reality."

What Jon was talking about is apparent already to my four-year-old grandson, who recognizes, when I show him, the separate reality of ants of any species (family Formicidae) as distinct from wasps of any species (family Vespidae). I think I'll have to wait a few years, however, before I explain to him the reality of their likeness within the order Hymenoptera.

"Look," Jon said, "there are only 7,000 families of all kinds of life on this planet. Of those 7,000 families, 105 are spiders. We have, right here in the museum, representatives of nearly all of them."

He is working hard to collect the rest.

"You know what I want to do?" Jon asked. "I want to create, as quickly as possible, a synoptic collection of the chunk of life for which I am responsible. Because two hundred years from now there's going to be a news conference held here, in the museum. There will be this minister of environmental affairs, who will announce that the Inventory of Life on Earth is complete, that there are 5,748,941 species of everything. And some reporter will stand up and say, 'But Madame Minister, didn't they say two hundred years ago that there were 12 million, 15 million, 30 million, even 80 million species?' And she'll answer in one of two ways. She'll say, 'Well, we lunched them all,' or she'll say 'They didn't know what they were doing two hundred years ago.' I want to make it impossible for her to give that second answer."

READINGS

Coddington, Jonathan. "Cladistics and Spider Classification: Araneomorph Phylogeny and the Monophyly of Orbweavers." *Acta Zoologica Fennica* 190 (1990).

———. "Designing and Testing Sampling Protocols to Estimate Biodiversity in Tropical Ecosystems." In *Unity of Evolutionary Biology: Proceedings of the Fourth International Congress of Systematic and Evolutionary Biology*. Portland, Ore.: Dioscorides Press, 1991.

———. "The Monophyletic Origin of the Orbweb." In *Spiders, Webs, Behavior and Evolution,* ed. W. A. Shear. Stanford, Calif.: Stanford University Press, 1986.

———. "Systematics and Evolution of Spiders." *Annual Review of Ecological Systematics* 22 (1991).

Hennig, Willi. *Phylogenetic Systematics.* Urbana: University of Illinois Press, 1979.

Johnson, L. A. S. "The Quest for an Optional Taxonomy." *Systematic Zoology* 19, no. 3 (1970).

Levi, Herbert W., and Lorna R. Levi. *Spiders and Their Kin*. New York: Golden Books, 1990.

O'Hara, Robert. "Representations of the Natural System in the Nineteenth Century." *Biology and Philosophy* 6, no. 2 (April 1991).

11

ROB WEST HEAVED HIMSELF up over the gunwale of the boat that carried his name. His father, Everett, who was steering, had named it the *Robbie Joe* when the twenty-five-year-old was an infant who napped on a spread-open life preserver on deck. Everett, a neighbor of mine in Maine, was still lobstering in those days and would take his son out when he checked his traps.

Rob took off his forty-pound lead weight belt and handed it up to his father as he prepared to climb aboard. It was past noon, and Rob said he'd had enough for the day; his air tank was empty, too. With only short breaks to change air tanks and grab a snack, Rob and his fellow diver, Travis Preston, had been in the water since seven in the morning, swimming constantly in weighted suits, battling underwater surges that threatened to topple them, and, with hand rakes, clawing sea urchins from the rocks. From time to time they had surfaced with bags full of sea urchins and handed them off to Everett, who gave them empty ones to fill again.

Everett dumped Rob's final mesh bag of sea urchins into an orange plastic tote that would hold more than sixty pounds of them when filled. But he didn't stop to sort them as he had been doing all morning. His face was marked with concern; he was scanning the water for the telltale bubbles that would show him where Travis, bigger, older, more experienced than Rob, but also wider-ranging, was still working under twenty-five feet of water.

The *Robbie Joe* is a thirty-six-foot snub boat, white with aqua trim, of the kind used up and down the coast for lobstering. For taking sea urchins it is fitted out with plastic sheeting stretched over a frame extending from the pilothouse. In wintertime the plastic serves as a windbreak and helps keep in the warmth from the portable heaters Everett carries on board. Even on a benign October morning a hot-water tank, heated by the boat's engine, simmered on board. The two divers had put their gloves and head-gear in the tank to warm them before putting them on. Maine waters are, after all, Maine waters and famously cold even in summer. The cold was just one of the things Everett worried about on his divers' behalf.

I'd been on the boat all morning, and although Everett was steadily busy, sorting sea urchins, slicing some open to check that they were fit for market, watching the screen of the bottom scanner, which showed the underwater profile so he could steer and keep the boat off hidden rocks, he was also constantly watching for the bubbles of his divers' breathing. He had no way to communicate with them directly, and he wanted to know where they were, that they were safe.

Everett's father and grandfather before him fished these same waters. He grew up on the water hearing all the old stories; he knew the dangers that lay under it and in recent years had learned how quickly even a simple swell could induce the bends. In the 1980s, when the market first opened up for sea urchins, and men along this coast started diving for them, there were two drownings. "They were a lot more casual about it in those days," Everett told me. "All anyone needed then was just a certificate that said he'd been to dive school. There wasn't any regular licensing. Nobody understood how dangerous it was. But even now these youngsters don't really understand. They take a lot of chances.

Guess I took chances, too, at their age. I worry about them all the time when they're under. I've taken ten years off my life with worry." A competent, reliable, enormously kind man, Everett appeared calm, but during the morning he worked his way steadily through a pack of cigarettes, betraying his anxiety.

We were out on the second morning of sea urchin season, October 7, and the weather couldn't have been better. The sun was shining, the waves were small, and the water, still comparatively warm from summer's heat, was quite clear. The season extends until April, and during the winter months, the job is much harder. Winter water is often choppy, riled, so murky that the divers can barely see. The waves obliterate the bubbles that come to the surface from their breathing, so Everett has no way to know where they are, no way to make sure that they are safe or that the boat, engines running, has not drifted over them. But worst of all in the winter is the cold. With the water temperature down to about 35 degrees Fahrenheit, diving gear can freeze up and a diver's body temperature drop dangerously. When the men get too cold, they surface, climb aboard, and pour water from the tank over themselves. "In the winter," Everett told me, "hypothermia is a real problem. I've seen my boys come up out of the water shaking and trembling with the cold."

At last Travis broke the surface and waved to Everett, who steered the *Robbie Joe* alongside, pulled up the mesh bag, took Travis's heavier fifty-five-pound lead belt and his scuba gear as Travis pulled himself up. Everett took down the flag that indicated he had divers overboard, sorted the last loads of sea urchins, and headed toward the dock.

Locally, taking sea urchins is called "egging." In the 1980s, as overfishing of some species of water creatures began to depress the incomes of fishermen along the coast of Maine, a good steady market for sea urchin roe opened up in Japan. The Japanese had just about depleted their own stocks of the prickly little animals, and the high prices they were paying had also nearly cleared them off the coast of western North America.

Sea urchin roe, as well as the urchins themselves, are called "eggs," but to be accurate, the prize is the animal's gonads, the reproductive tissue of either male or female before eggs or sperm appear. Once eggs and sperm are produced, later in the sea urchin's annual cycle, the gonads are no longer marketable.

Sea urchins are relatives of sea stars, sand dollars, and sea cucumbers. The species off the coast of Maine, the one that Everett and his divers were taking, is the green urchin, *Strongylocentrotus droebachiensis,* a name which, spelled out on my typewriter, is as long as the diameter of a marketable specimen. "Urchin" is an old English name for the prickly hedgehog. In size and appearance it looks very like the burr of a horse chestnut, slightly flattened on the bottom.

Some of the animals that had made me feel so uneasily alien in my brief underwater venture with Klaus Ruetzler off the coast of Belize were sea urchins of various shapes and species. I had found the waving of their spines in the gentle currents not threatening

A green sea urchin, *Strongylocentrotus droebachiensis,*
tilted to show its mouth (½ lifesize).

but distinctly unsettling. I'd hugged my stunned curiosity about them to myself as I sat drinking Klaus's restorative rum that day, and now, after many months, had brought it to Everett West's boat on a bright Maine autumn day.

I noticed that these green sea urchins waved their spines, too, as Everett poured the contents of a mesh bag into a plastic tote. They must have been alarmed at leaving the water. This species often uses its spines to arrange bits of weed and detritus to cover and hide itself. Taking urchins out of the water does not kill them, and in Everett's refrigerated truck, they and the others his wife and daughter were buying from other eggers would still be alive when he hauled them to the processing plant in the evening. And within two or three days the roe would be on tables in Japan.

We had left a dock off the road Everett and I share at seven that morning. We'd picked up Rob and Travis at the next dock down the peninsula and headed out to sea. For several years I'd watched fishing boats go out in the early morning as I drank my coffee. I'd watched them disappear along the curve of the peninsula toward open water. And now I was on one of them. The autumnal maples and birches lit up the shore, red and yellow and orange, glowing in the low morning sunlight as we headed out.

Beyond the peninsula's end, in the shelter of one of the big islands, Travis and Rob had flipped expertly backward from a sitting position and into the water for their first dive. As Everett sampled the urchins they brought up, slicing them open on the gunwale with a big knife, he became more and more dissatisfied. Under the prickly spines, within the urchin's protective hard skeleton — it is called a test, not a shell — are five cavities, which by October should be filled with fine-grained golden roe. But the color and amount were not up to Everett's standards.

Commerce ranks the best-colored roe as grade A. But the ones Everett cracked open were grade B or less, he said. And many were under the regulation two-inch-diameter size. He threw back most of the haul and told the divers we would move back toward the mainland. We ended up just off the rocky town park on the peninsula, where I'd first explored the tide pools. The water deepens quickly there, and the divers were soon working at nearly twenty-five feet. Everett cracked open samples from the first bags they brought up and was satisfied with them. He pointed out the roe-stuffed cavities to me and said that the rich yellow color was much better and the finer grain superior, making them grade A. He nudged some toward me with the point of his knife and suggested I taste it.

I hesitated for a moment. This animal had just been waving its spines at me. But its severed halves had just been tossed overboard and the roe would follow if I didn't take it, so I fingered up a portion of grade-A sea urchin gonad and ate it. It tasted of the sea, not salty or fishy, just essence of ocean, surprisingly sweet, breeze-fresh, delicate. The Japanese eat it on happy, celebratory occasions, but they also think that it guards the body against dementia and cancer.

It seemed for a while in recent years that golden sea urchin roe would be economic gold for impoverished Maine coastal communities. The *Robbie Joe* that day carried back about 1,200 pounds of sea urchins. In years past the processors usually paid the eggers more than a dollar a pound. There were so many urchins in those days, even in shallow waters, that many of the fishermen started taking them. And in the winter months, just as the lobstermen were storing their traps, the sea urchins were stuffed with yellow roe as they built up for spawning in the spring. Scallops had long been taken from these waters by draggers, boats with heavy weighted nets towed behind, and at first that was the way sea urchins were taken, too. But although some are still dragged, the fishermen soon discovered that scuba divers can be more discriminating, can pick out the ones that appear to have a larger quantity of roe, and can leave in place for growth urchins too small to be salable. Gradually the majority of eggers began working with a dive crew, as Everett does.

Everett told me that he began dragging for sea urchins eighteen years ago while he was still lobstering. But he soon switched to using divers and then, about seven years ago, he gave up lobstering entirely.

Sea urchins have been taken in small numbers from the coastal waters of Maine for many years; the state first started keeping records on landings back in 1929, when 3,000 pounds were hauled in and sold in specialty shops in Boston and New York. Their sale was a very minor part of the income of Maine fishermen. But by the 1980s the big ground big fish — haddock, cod, and the like — had been overfished for years. And more and more lobstermen were finding smaller and smaller lobsters for both the increasing domestic market and the international one that air freight was creating. So when Japanese traders, with the yen strong, began exploring the possibility of buying sea urchins in Maine, egging looked like it might save the fishermen and their towns.

By 1987 all trade arrangements had been made and the processing plants built. That year, more than 144,000 pounds of urchins were landed, bringing a modest $250,000 into the Maine economy. But every year thereafter the landings, and their value, increased. In 1993 the landings were more than 41 million pounds, representing revenues of $26 million.

But in the years that followed, catches dropped and the eggers had to spend more hours out in the water to bring them in. In 1996 the landings amounted to just about half of what they had been in 1993. At the same time the yen was slipping in value and the Japanese economy was not as strong. Since sea urchin roe is a pricy luxury food in Japan, the market for them there was beginning to soften. Everett told me that on the day before, the first buying day of the new season, the catch brought the same as in years past, but on the day that I was with him, it would drop to ninety cents and would probably go down further in the days to follow.

The catch was also declining. "I've seen it go down every year," he said. "And the quality of the roe hasn't been as good either."

What happened?

Eggers and biologists agreed that more and more people were going after less and less resource. But, as might be expected when livelihoods are at stake, there was considerable disagreement about what can be done to save that resource, if it can be saved at all, and what the government's regulatory role should be. The biology of the sea urchin had suddenly become very important.

Ted Creaser, of Maine's Department of Marine Resources, was the man who gave me the figures for the sea urchin landings over the years and their value. I told him that I'd spent a day talking with Bob Vadas about sea urchin biology. "Ah, yes, Bob Vadas," Ted said. "There's no one who knows more about sea urchin biology than Bob Vadas."

I'd spent a summer day with Bob in his office at the University of Maine in Orono. Back in the 1960s, when Bob got his Ph.D at the University of Washington, he was a botanist. He soon discovered that to understand ocean algae, which was what interested him, he had to understand sea urchins that grazed upon it. As he described it, he became one of the early "cross-over biologists, putting botany and zoology together."

The sea urchins I'd seen off the coast of Belize belong to a variety of tropical species that, because of what they eat, are important to the health of coral reefs. The dominant urchin there, and also the biggest and showiest of those tropical species, is *Diadema antillarum,* the long-spined black urchin, with spines that measure up to six inches in length. They move in and around the reefs, feeding on the algae that grow on them, scraping them clean as they feed. Those algae-free spots are the places where larval corals (and sponges, too) can then attach themselves and continue the process of building up the reef, home to so many other animals.

In the 1980s, a disease specific to that species and no other, black urchin plague, spread quickly throughout the Caribbean and killed very nearly all of them. Scientists found that the fish and other animals which had fed upon these urchins simply switched to other species; they also found that the algae grew thicker without them and that a few small fish benefited from its cover. Initially it seemed that the black urchin might be a redundant species, a "spare part," of the sort the Swedish researchers mentioned in Chapter 6 were writing about. But that turned out not to be true. In time it became apparent that without the long-spined black sea urchins to clear away the algae from the reefs, the larval corals had no place to make their adult homes and, as a result, the corals began to deteriorate.

Maine has no coral reefs, of course, but the interaction between its green sea urchins and the rest of life underwater, as well as with human life and economy above it, is enormously complex.

For 150 years there have been reports of periodic buildups of green sea urchin populations along the eastern coast of North America, but only recently has there been much research on the factors that cause and constrain those buildups. In 1847 a scientist first recorded that fishermen were reporting increasing numbers of urchins along the coast. Then in 1868, Sir John William Dawson, the Canadian geologist for whom the Yukon gold town was named, published a paper on what sea urchins eat. "The nature of its food," he wrote, "does not seem to be generally known. In dissecting some specimens . . . collected last summer I found the intestine full of sea-weeds. It would thus appear that the curious apparatus of jaws and teeth possessed by this creature is used in a kind of browsing or grazing process. . . . The sea urchin is thus a kind of submarine rodent."

More than one hundred years later, Bob Vadas wrote, "Sea urchins are simple animals, consisting of gonads and gut surrounded by a calcareous test." It is a simple animal, but it is an exceedingly curious one, beginning with those jaws that Dawson mentioned. They are a five-part affair of skeletal rods with powerful muscles

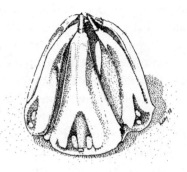

Aristotle's lantern (enlarged many times),
inverted to show the teeth.

that open and close the five teeth rather in the manner of a drill chuck. The teeth grow continually as the animal grinds them down in feeding.

The mouthparts have an unusual name, "Aristotle's lantern," taken from a passage in Aristotle's lecture notes on animals, circa 322 B.C. He said that the urchin's "so-called eggs were used at table . . . and some use it as a remedy for strangury [slow and painful urination]." He then went on to give an uncharacteristically confused description of the animal's physical appearance (perhaps the transcriber of the lecture was at fault), ending, lamely, by saying, "In respect of its beginning and end the body of the urchin is continuous, though in respect of its superficial appearance it is not continuous, but similar to a lantern lacking its surrounding skin."

The translation I am using here is the standard one by A. L. Peck from the Loeb Classical Library, but the passage is confusing even in the original, apparently. Translators and historians have argued over it for centuries. A sixteenth-century French medical man, Guillaume Rondelet, a notorious purveyor of misinformation (we will meet him again in another context) decided that Aristotle was speaking of the sea urchin's mouth and five-part jaws, which, with their supporting structure, are rather lantern-shaped. He therefore called that part "Aristotle's lantern," a name so unusual that it stuck, even though other translators and historians have insisted that Aristotle was describing the whole animal. Some went so far as to place a light inside a sea urchin shell to prove that it could make a creditable, if tiny, lantern. Such are the ways of scholarship.

Well, a sea urchin *is* a hard animal to describe to someone who has never seen one, despite its simplicity. For one thing, it is radially symmetrical. We are bilaterally symmetrical animals, which means that if you draw a line from the top of our heads down the center of us to our feet, each half mirrors the other half. With a radially symmetrical animal, the likeness is not just on either side of a center line, but all around it. Aristotle's lantern — the mouthparts — is one pole of that line, the center of the urchin's flattened bottom. ("The reason being," Aristotle sensibly said, "is that they obtain their food from below.") The line's other pole, on top, is the urchin's anus. Inside, the two poles are connected by the gut, which threads its way amid the five chambers where the roe forms. In thinking about this animal, it is helpful to keep the number five in mind. Everything to do with a sea urchin comes in fives. The globe of the urchin's body is made up of rigid plates, in multiples of five, which bear the spines and little tube feet for walking. The plates are perforated with openings through which the urchin sheds eggs or sperm in spawning season.

The five rows of tube feet correspond to the tube feet on the underside of a sea star, one of the clues, along with the radial symmetry, that the two animals belong to the same phylum, Echinodermata (the spiny-skinned). And, in fact, one way to visualize a sea urchin is to imagine a sea star with its five legs pulled upward until they meet. Add spines to that picture and you have a sea urchin.

"The sea urchins," Aristotle said, sounding surprised, "have no fleshy part . . . they are all completely empty." The gonads serve, in a way, as "fleshy" parts, because they store whatever food energy the animal acquires in feeding. The gonads swell when the animal eats high-protein feed, such as the kelp it favors. In times of famine, the urchin can draw nutrition from the gonads, which shrink in size and can even be completely reabsorbed. State regulations and commercial processing value dictate that harvested sea urchins must contain at least 10 percent roe by body weight. The one that Everett had declared grade A by color and texture of the roe had, he judged, 13 percent.

When urchins are feeding well, the gonads increase in size throughout the winter months. Then, between February and May, along the Atlantic coast, they spawn: through those openings in the five plates, males and females shed eggs and sperm into the water, and when they meet, tiny larvae are formed. The sperm are active for only a few minutes in the water, so reproductive success depends on thick aggregations of sea urchins all spawning simultaneously. If eggers take urchins in great numbers, sperm and eggs will have a harder time meeting.

Bob Vadas took me into his laboratory and showed me vials of seawater cloudy with sea urchin eggs. He captures sexually mature urchins, puts their sperm together with the eggs, and raises larvae. The Japanese are already raising urchins in this manner on a commercial basis, and returning lab-raised larvae to the ocean may turn out to be the most economical way to restore the populations along the coast of Maine.

In the cold springtime waters, the nearly transparent, asymmetrical larval sea urchins float with the plankton, feeding and growing for six to eight weeks. Then, still only one-thirty-second of an inch in size, they sink to the bottom, transforming themselves in less than a minute to tiny, round, symmetrical urchins. They then begin to scrape and eat just as the new spring kelp begins growing. During the next two to four years, their body cavities are usually empty, or nearly so, for they need all the food energy they can get to grow and mature. One of the reasons that the stock of urchins is declining, researchers believe, is that too many small, immature animals were taken during the 1980s, depleting the future breeding stock.

Today buyers are supposed to reject any egger's catch if more than 2 percent of the individuals in it are under two inches. Everett kept his two-inch measurer close to hand to check any that looked too small to his trained eye.

Bob Vadas had told me that pressure from the Japanese market to meet the demand during summer holidays, when sea urchins are traditionally served, had pushed researchers to see if the breeding cycle could be altered to make them available then. From last June until August, his researchers fed confined sea urchins on a high-protein algae, and they discovered that good-quality roe quickly developed on the diet, even though the urchins' cycle was six months out of its natural phase at this latitude.

Everett knew about the sea urchins' need for high-protein grazing. He'd looked disappointed as he dumped one string bag full of extremely clean-looking urchins into a plastic tote. "I know where these eggs came from," he said. "Don't have to be underwater to tell Rob's been working the ledge. You can see there aren't any pieces of kelp in these. The eggs won't be any good because they don't have anything to feed on there." Next time Rob surfaced, Everett showed him some cracked-open sea urchins from the previous catch with little or no roe in them and told him to stay away from the ledge, to go back out to the beds of kelp.

After Rob went back down, Everett went on to talk about the sea urchins' diet. "Seems like they can live a long time without much to eat. Once I found one that had chewed its way into a soda can and was just living inside it with nothing at all." Bob Vadas had told me they gnawed away at almost anything, including tin cans.

Lacking high-protein algae or soda cans, sea urchins eat a variety of things: marine worms, snails, barnacles, mollusks, hydroids, wood, decaying plants, and sponges. Bob had recently completed research showing that when a green sea urchin is injured, others in the group retreat rapidly, but within half an hour will return to eat the wounded animal. They can even eat sewage and can tolerate large amounts of chlorinated hydrocarbons in it.

There aren't good census data on urchin populations, but around the time people began to talk about lobsters being over-fished, there seemed to be a lot more sea urchins around than there had been: another one of those big population bulges had come along. The fishermen insisted that lobsters didn't feed on sea urchins, but a scientific study, which Bob said was flawed, found otherwise, and the idea took hold in scientific circles that the increase in sea urchins proved the decline in lobster populations. "Despite the fact that later studies have shown it isn't true, people still talk about lobsters as a keystone predator feeding on sea urchins; you even find it in textbooks. The notion got a lot of PR," Bob told me. His own studies have shown that lobsters don't even like sea urchins as food and avoid them when possible.

It may be that any cyclical growth and decline in sea urchin populations is in response to natural, biological rhythms and that fishermen and their catches have nothing to do with it. Even Bob Vadas doesn't know for sure what caused the big recent buildup. "Although," he said, "one of the contributing factors may be that sea urchins have acute chemical sensitivities. My research shows that they don't aggregate specifically to one another, but they can tell where the good food is, so a rich source of food attracts masses of them."

And although Bob's research disproved the notion that abundant lobsters kept the sea urchins in check, he has also shown that predation by big fish such as haddock, cod, and wolffish may have limited the urchin populations. Those fish were once common in Maine coastal waters, but they were taken in such numbers that they are seldom seen there today. However, they can still be found farther out at sea, and Bob reasoned that conditions there might be similar to past conditions along the coast. So he set up an experiment. He took his research team sixty-five miles out to sea near some offshore islands. Although they found veritable forests of kelp there, the urchins' favorite food, the researchers found few urchins, and those they found were small. The big predatory fish lived there in goodly numbers, and within their guts, the researchers found big sea urchins. In experiments with tethered sea urchins, they found that the predatory fish quickly ate large ones but often ignored small ones. "Overall," Bob and his coauthor, Robert Stenneck, concluded, "these data suggest that the structure of shallow nearshore marine communities in the western North Atlantic was controlled by a suite of large predatory fish species. . . . In the absence or reduction of large fish, such as that resulting from overfishing . . . we would predict a shift in the structure of the community from one controlled by fish and dominated by kelp to one controlled by sea urchins."

Bob Vadas is one of two scientists who serve on an advisory panel on the sea urchin fishery, along with eggers and processors. At one meeting a new member asked him, "What are we saving them for?" Bob sighed when he told the story.

That question has both an easy and a hard answer.

The easy part is that the industry, following the regulatory agencies' euphemism-generators, calls the taking of sea urchins "harvesting." It is a term I've avoided using because it makes me think of soybeans, but you can see how a fisherman would regard a resource that is called "harvestable" in the same way a farmer would regard a crop he has planted. And if someone is speaking in that misguided way, the best answer would be that a harvester ought not to consume his seed stock. A very small knowledge of the life cycle of sea urchins would show that theirs is not an annual birth-growth-reproduction-death circle like that of a soybean. Urchins take several years to reach sexual maturity and reproduce, and, because of the peculiarities of their reproduction, a rather thick mass of them must be in one place for success in making baby sea urchins. We don't really know how large the population must be, but it is in the fishermen's self-interest to leave a goodly stock in place if they want to continue making a living at egging.

It is more difficult to make a case that a particular total number of sea urchins is needed to make a "normal" population. Bob Vadas's research offshore suggested, but did not prove, that recent big aggregations a few years ago close to shore might be the result of the near elimination in those waters of some formidable fish predators, such as cod and haddock, through overfishing. There is no good census data, only people's vague belief that within the remembered past, before the commercial taking of sea urchins began, there were larger numbers than in the distant past. So it is possible that the buildup was the result of overfishing, but it is also possible that it was part of some unknown biological cycle. Or it may be that there never was a buildup at all, so limited is our knowledge of the ways of sea urchins. An analogy can be made here to the question of what the effect humanity has had on populations of North American native bees. We can't know for sure because we don't know what their populations were before people came to this continent.

However, there is a more basic issue inherent in the question the fisherman posed to Bob Vadas, one that is frequently asked these days as the needs of a growing human population come up against the needs of other species. Bob's questioner could have said, more bluntly, "Look, I don't care if they are seed stock, I've got a family to support and right now I have no other way to make a living. Why can't we kill all of the sea urchins and then find something else to fish for? Why can't every last one of them be eaten? What good are they, anyhow?" That cluster of questions represents a variation upon the devil's question I asked of Peter

Larsen and Hank Tyler and, by implication, lots of other zoologists I've talked with during the course of my writing life:
Why should we pay attention to strange little animals with funny
names? What are we saving them for? In this basic sense, I think it
is a perfectly good question. I think it deserves answers.

I've put that question to others in different forms, but it is only
fair that I, too, should have to account for my interests and not
simply join Bob Vadas in his sigh.

My answer has two parts, one soft and one hard.

The soft part is that once we make a pet of an animal, human
beings are loathe to run it off the planet. Today, in the United
States, I think I am safe in asserting that it would be impossible to
begin unlimited harvesting of dogs and cats even if the price were
right. We have too many beloved kitties and dear doggies in our
homes, and we understand them too well, even though the fond-
est of us would have to admit that economically our pets are use-
less, even a financial drain. They mean something to us; we *know*
dogs and cats. A knowledge of the biology and behavior of animals, even the ones we considered icky before we knew them —
wasps or jellyfish, for instance — turns them into intellectual pets,
as it were, and makes us feel kindly toward them. That may be a
soft, emotional response, but it is characteristic of our species (as
characteristic as is the cruelty that comes from a lack of knowledge) and is not anything to be ashamed of. On the contrary: it is
called compassion.

The hard, steely-eyed part of my answer is nearer to self-interest
and therefore, in some ways, should be more compelling. It has to
do with one of my lifelong interests, which is the theme of this
book: time and its effects on biological processes.

The case of sea urchins is a good and tough one to examine because they, like us, may be more expendable than some of the
other animals that have shown up on these pages, whose rights to
existence would be easier to establish.

Relative to a number of other presently living invertebrates, sea urchins are a comparatively young group. And unlike a number of groups, because of their hard, shell-like casing, we have a fair fossil record of them. Modern sea urchins, of which there are today something like a thousand known species, begin to show up in that record in the Triassic, a mere 240 million years ago. By that time there were already dinosaurs, turtles, and trees that would be recognizable to us today, as well as some small scurrying mammals. But those last named still had more than 200 million years to go before any group of them evolved into anything we'd be proud to claim as a relative and began using clever fingers and a brain in ways we would appreciate.

So, even allowing that they are considerably younger than sponges or worms, let alone bacteria or the other staples of the earth's biota, the sea urchins have been more tested by changing world conditions than we have. (The dinosaurs failed the test.) When our particular species first appeared, the sea urchins were here, as were a lot of other invertebrates, including all those wildly successful insects, and modern reptiles, plants with flowers, monkeys, birds, bats, fishes, and many, many creatures with shells. Although the weather continued to fluctuate, the world we were born into has remained quite stable during our species' lifetime. When we came about, our planet was thus and so; the animal life was thus and so; the plant life was thus and so. And all of it had spent a great deal more time than we have lived since in working out complex interrelationships that, taken all together, have created our ecosystem. This is pleasant for us, and we have thrived in this ecosystem.

Life, as a generality, may be adventuresome, highly adaptable, going in for wild experiments and speciation to take advantage of new niches created by changing conditions. But particular individuals and their species, as Loren Eisely once pointed out, are always conservatives. We are biological Republicans the instant we are born: we come into a world constituted thus and so, and our tolerance for variations — even allowing for some human technological fixes — doesn't allow much leeway about the precision of that thus and so — earth, ocean, and atmosphere chemistry as well as the biotic foundation on which we depend. As individuals and as a species, we want everything to stay the Good Old Days. If there were such a thing as Species Consciousness, it would say, "Like it just fine the way it is. Don't change a thing."

Our species is a particularly new and raw one. It has never been pruned by world cataclysms like those that ended the dinosaurs or the trilobites or the majority of animals that life has played with during the billions of years it has been present on the planet. Some kinds of life did survive cataclysms, notably the sponges and worms. Even the sea urchins have been seasoned a little. If the past is any guide, there will be future cataclysmic events in one part of the world or another, and they will take out the tenderest parts of the biota. Then, as Klaus Ruetzler put it in another context, "We would have a different ecosystem."

Can we survive such a catastrophe? Can we live in another ecosystem? There is no way of knowing, and when such a catastrophe occurs, our species may already have passed out of existence for other reasons. But the odds of survival are not in our favor. Again, if the past is a guide, and it is the only one we have, those species best able to survive severe environmental events are those that have been through earlier ones. And we haven't. Our extinction would be bad for us, but from a planetary point of view it would make no difference except to a few of our fellow travelers — crab grass, city pigeons, and suchlike. According to what we know, we seem to be one of those redundant species.

On the other hand, a great many of the organisms that were present in the world when we came into it may be much more important to us than we are to them. The plants and animals and bacteria and molds and viruses that make up our ecosystem, many of which are invisible to us as we walk through this world acting as though we own it, were already in place when we came along. We don't know very much about our ecosystem or about them. Western science is only a few hundred years old, and although we may think we know more about all of it than our ancestors did, we know very little. We don't even know the identities of a great many pieces of this ecosystem and practically nothing about how the totality works.

Particular animals, the sea urchins, for instance, are also fairly new and possibly redundant. Maybe we could harvest every last green sea urchin and the ecosystem would be just fine. But we didn't think that the long-spined black urchin was very important until the black urchin plague struck in the Caribbean and created an ecological experiment that showed us that these animals had a significant role in the health of coral reefs, on which much other life depends.

But let's grant that we can be profligate with the green sea urchins. It is quite possible that their loss wouldn't hurt us. Maybe we can be profligate with spotted owls, too. After all, we haven't noticed any problems caused by the absence of the ivory-billed woodpeckers and passenger pigeons. Maybe we won't miss one whit the golden toads or the Karner or Xerces blue butterflies or any of the other specific animals with funny names on which we have put intolerable pressures. But eventually, if we continue to be profligate with the biota and all it needs, we will, of a certainty, cross some line that separates the ecosystem in which we have thrived from a new one. Crossing the line will represent a test to us, a test that the odds say we will flunk. We don't know where that line is, and we can't begin to understand what it would mean to live in a new ecosystem because we don't understand the one we have. That makes me nervous.

So call this a conservation ethic based on nervousness, but I think the smart money would agree: be really, really careful with all the things in the world, particularly those that were here before we were. We may need them more than they need us. Best get to know them. If we enjoy living in this world — and I do — and want our descendants to enjoy living in it, too — and I do — we should consider that nervousness tempered with compassion is not a bad conservation ethic.

Bob Vadas believes there are several ways to create what he calls a sustainable fishery of sea urchins, which would bring income to coastal Maine and not reduce the resource. Most simply, he told me, sea urchins could be taken from a place where they are too many to places where they are too few. Commercially raised larvae could be released into open water. And commercial aquaculture could supplement the taking of free-ranging sea urchins. In the meantime the State of Maine has set up zones for taking sea urchins with different dates of seasonal opening and closing to reflect latitudinal variations in sexual maturity. And within those seasons, the regulators have blacked out certain days when egging is not allowed at all.

Some people who object mightily to any regulations whatsoever would agree with the man who asked Bob Vadas, "What are we saving them for?" Everett is not one of these people; he is a thoughtful man and has seen the decline in number and quality of the takings. But he doesn't think the current regulations work very well. "In the first place, they came too late. They ought to have thought this out a lot earlier. And I don't like all those blackout days in the season. I won't go out in bad weather, but some feel they are forced to go out in weather they shouldn't, and that puts divers in danger. Mother Nature regulates us pretty well during the winter just as it stands." Anyone who has ever spent winter along the coast of Maine will understand what he means. It is a blustery time and even a squall — let alone a heavy storm — keeps anyone with good sense far from a small boat on choppy, dangerous seas, and by his fireside — whether the squall comes on a permitted egging day or a blackout day. And egging is a winter fishery.

"It would be better," Everett said, "to drop those blackout days and just regulate the catch." Everett believes that allowing 2 percent of the catch to be undersized is too lenient, that no urchins less than two inches should be taken. He'd also like to see all boats restricted to carrying only two divers — he himself sometimes takes more men later in the season — and he thinks each boat should be held to a ten-tote limit per diver each day. License enforcement should be stricter, too, so that eggers from one zone stayed out of the other.

In the weeks that followed my trip with Everett, Rob, and Travis, eggers all along the coast reported fewer sea urchins out there for the taking. But then the winter brought the crash of Asian economies and the weakening of Japan's. In addition, Russian fishermen, nearer to market, began offering at a lower price a species the Japanese preferred to the green sea urchin. A good deal of the Maine sea urchin processors' trade faded away. Prices paid to fishermen dropped until it was hard to make a profit taking a boat full of divers out into frigid waters, even on days that weren't blacked out. Boats stayed on their buoys and sales of stored urchins dropped to new lows. During this lull the sea urchin population may build a little. But the question of "What are we saving them for?" may be asked of them again. And it will certainly be asked of other species.

On that day in October, during our trip back to the dock, I asked Travis and Rob what they did when they weren't diving. Rob digs blood worms for bait and cuts spruce boughs for the local wreath-making factory in the winter. Travis's other job is laying linoleum. The two young men were peeling off their wet suits and putting their clothes to rights as we talked. And, being young, their tired muscles seemed to recover at a rate in direct proportion to the lessening distance from the cheeseburgers they were planning to order for lunch, which they described to each other in luscious detail.

Travis, who had been bantering and playing the clown all morning, grew serious when he talked about why he dives. Some years earlier he'd had a bad diving experience, had lain unconscious under water for ten minutes and been declared drowned. So he has every right to fear it, but he doesn't. "It took me a while to get back in the water, but I did. And, you know, I like it down there. Oh sure, the money from egging is good, but the real reason I do it is because you see so many interesting animals and plants down there. Some of them are luminous, and where it is dark they glow. There's a lot of different colors in the lighter places, the rocks, the kelp. It's beautiful underwater."

READINGS

Aristotle. *History of Animals.* Trans. A. L. Peck. Cambridge, Mass.: Loeb Classical Library, Harvard University Press, 1965.
Chenweth, Stanley. *The Green Sea Urchin in Maine.* West Boothbay Harbor, Me.: Maine Department of Marine Resources, Benthic/Demersal Division, n.d.
Creaser, Theodore. *Data Sheet on Historical Landings of Sea Urchins.* West Boothbay Harbor, Me.: Maine Department of Marine Resources, Marine Fisheries Laboratory, 1997.
Dawson, J. W. "The Food of the Common Sea Urchin." *American Naturalist* 1 (1868).
Elner, R. W., and R. L. Vadas. "Inferences in Ecology: The Sea Urchin Phenomenon in the North Atlantic." *American Naturalist* 136, no. 1 (1989).
Hendler, Gordon, et al. *Sea Stars, Sea Urchins, and Allies.* Washington, D.C.: Smithsonian Institution, 1995.
Lawrence, John M. "On the Relationships between Marine Plants and Sea Urchins." In *Annual Review of Oceanography and Marine Biology.* London: Allen & Unwin, 1975.
Lessios, H. A. "Mass Mortality of *Diadema antillarum* in the Caribbean: What Have We Learned?" *Review of Ecological Systemss* 19 (1988).
Lessios, H. A., et al. "Mass Mortalities of Coral Reef Animals." *Science,* November 1983.

Paine, Robert, and Robert L. Vadas. "The Effects of Grazing by Sea Urchins *Strongylocentrotus spp* on Benthic Algal Populations." *Liminology & Oceanography* 14 (1969).

Vadas, Robert. "Comparative Foraging Behavior of Tropical and Boreal Sea Urchins." In *Behavioral Mechanisms of Food Selection,* ed. R. N. Hughes. Berlin: Springer-Verlag, 1990.

————. "Experimental Evaluation of Aggregation Behavior in the Sea Urchin, *Strongylocentrotus droebachiensis." Marine Biology* 90 (1986).

Vadas, Robert, and Robert S. Stenneck. "Overfishing and Inferences in Kelp–Sea Urchin Interactions." In *Ecology of Fjords and Coastal Waters,* ed. H. R. Skjodal et al. Amsterdam: Elsevier Science B.V.

Vadas, Robert, et al. "Protection of Sea Urchin Roe: Rapid Summer Enhancement in Green Sea Urchins." In *Proceedings of Information Session, Strongylocentrotus droebachiensis.* Moncton, N.B.: Shippagan, N.B., Centre de recherche et de développement des produits margins, in press.

12

BEYOND THE HOUSE on my farm in Missouri was a field I kept open with an annual brushhogging, and around it were wild woods. Sometimes I would walk out there on clear moonless nights. The air was transparent; there was no light pollution, for the nearest town was miles away. The stars were many and sharply bright. I was alone with them, surrounded by the darker shapes of the encircling trees, and I felt as though I were at the center of the universe. It was a fine place to view visiting comets, figure out the constellations, look through binoculars at the Andromeda galaxy, contemplate its possible collision with ours 5 billion years hence.

Occasionally, in late summer or early autumn, out by the old pond in the middle of the field, the stars' twinkle and shine in the sky seemed to be answered by shine and glow on the ground. The grass all around the path to the pond's edge would be dotted with lights. The first time I noticed, I remembered that years before, while traveling through Tennessee with a friend, we'd seen the same thing and wondered if there were glowworms in this country. I read, after that trip, that glowworms were indeed Old World beetles, the females of which never grow wings even after they mature sexually. What we did have in this country are various species of firefly larvae that glow, particularly when disturbed. Fireflies are also beetles. They contain an unusual chemical that makes them taste just awful to predators; as a result they have few natural enemies.

One Missouri evening when there was shine above and shine below, I brought a flashlight with me, aimed it at the lights in the grass, and scooped up a couple of the glowing creatures. They did indeed look like beetle larvae, and when I compared them with the firefly larvae pictured in one of my beetle books, they seemed the same. They were dark in color with a hint of spots, and underneath their nether ends I could see the glowing parts that the book called their "lantern." After looking them over, I took them back out to the grassy bank of the pond and let them go. They were shining furiously; I suppose they were annoyed at being handled. I saw their kind many times afterward and always wondered to what purpose they glowed. I knew that adult fireflies, which crowded the velvety summer Ozark nights, used lights as signals in their matings, but these larvae were not sexually matured.

Washington has a lot of fireflies, too, but even in our quiet residential neighborhood it is never really dark at night. Streetlights, yard lights, house lights, and the ever-present glow from downtown keep it merely dusky. But one night in early September a storm knocked down the power lines, and the streets around our house were dark. I took the dogs out for their walk before bedtime, and we found our way by the light of dim city stars. After circling a couple of blocks, we headed back toward the house. Although our front lawn had appeared perfectly ordinary when we left, upon our return it was shining with specks of light far brighter than the hazy stars overhead. Ours was the only neighborhood lawn so decorated, and I stood in the middle of it, watching the steady glow of what looked to be every ninth blade of grass; lights were massed so thickly under the azalea bushes that the twigs above them were illuminated. I was reminded of a Christmas Eve, years before, when I was camping with friends high in the hills surrounding Mexico City and saw luminarias shining everywhere below us.

At last I went into the house and found my way to bed by lantern light. My lantern was fueled by kerosene, and it and the light bulbs that came back on in the middle of the night are crude by comparison to those firefly lanterns. Any light we contrive is prodigal, merely the byproduct of heat. But fireflies and a lot of other animals make a heatless light simply by being alive.

"Living light" is a perfectly respectable scientific term, sometimes also called bioluminescence. I found that a lot of people were studying it and thinking about it, and had been over the centuries. The computer catalog at the Library of Congress listed seventy-one monographs on the subject, about half of which were in Russian or another Slavic language. I gave up counting the number of scientific papers.

Humans can glow under special circumstances. The blood of smokers is weakly chemoluminescent, and I read reports in a nineteenth-century text that dying people sometimes shine. Modern researchers have found that several mortal conditions make human blood give off even more light than that of smokers, so that may be what is behind those old reports. But in general we can't produce living light. An enormous number of other organisms — plants, animals, bacteria — can, however. Perhaps that is why we are so fascinated by the light; perhaps that is why so many researchers work on the subject.

Thomas L. Phipson, the writer from whom I learned about the glowing of moribund people, also wrote that in Italy, where true glowworms live, it had once been the custom for 'youths to decorate the hair of their mistresses with those 'diamonds of the night,' which, the author added unromantically, "were probably less expensive than a pearl necklace." He also included a report of a letter he had received in 1858 from a French correspondent, one

M. Adrien of Pont Saint-Esprit. I have no choice but to give the passage in full.

"One summer's night after a rainy day," says the writer, "I saw the ground sparkling with a whitish phosphoric light whilst sprinkled with warm urine, and I recognized at the same time the presence of numerous small worms. . . . The phenomenon was so curious that I took up some of these worms and carried them into the house to examine by the light of my lamp. I immediately recognized them to be small *Lumbrics*. . . . Returning to my garden, with a lantern, I saw at the same place many *Lumbrics* crawling on the ground with their usual slow and regular mode of progression. But they showed no light; and when the lantern was put out, their presence could not be recognized. But as soon as they were in their turn sprinkled with warm urine, the phosphorescence of their entire bodies shone forth and illuminated their wriggling movements." The writer of this letter says he has since repeated the experiment many times.

Toward the end of my talk with Sam James, the earthworm specialist, I finally blurted out my question. "If a Frenchman were to pee on a nightcrawler, would it glow?"

Sam laughed and said, "It might. Bioluminescence is common among earthworms, particularly when they are agitated. Perhaps it is protective for them to give off blue-green light when they are disturbed; maybe it distracts a predator long enough for an earthworm to get away. But no one really knows. No work has been done on it."

So. Earthworms do it. Fireflies do it. A few other insects and millipedes do it. And they are all animals that are active in dark places or during dark hours. Living light cannot compete with that of the sun, so animals that are active in daylight don't produce their own light. Some make use of sunlight to illuminate color and pattern to protect themselves or alarm their enemies. Some use iridescence, some use reflected light. For example, one ocean-dwelling water strider gathers with others of its kind when predators are about; they move in rapid, nervous dashes, and the light reflecting off their grouped bodies as they zig and zag has been shown to disorient and confuse their enemies.

I remember being completely bamboozled once when I was river rafting in Costa Rica and saw my very first space alien float gently over my head in a tiny, shimmering violet flying saucer. It was *Morpho cypris,* a tropical butterfly performing the magic trick it does so easily every time it takes wing. At rest it is a gorgeous creature with iridescent purple-blue scales on its wings, and when it flies it does so with very slow wingbeats which, combined with the iridescence, create the optical illusion that it is globe-shaped and floating. As a human I was fooled, didn't even realize it was a butterfly. If I'd been a predator I probably would have been fooled, too, wouldn't have known where to sink my beak into substance as the orb lazed by.

Such uses of existing light are very common among terrestrial animals, but genuine living light, bioluminescence, is not, fireflies and earthworms notwithstanding. When Travis, the sea urchin diver, told me about how beautiful and interesting it was underwater, he mentioned the luminosity. The real experts of living light are in the sea.

Some animals that live near the surface, where light penetrates a little, have glowing undersides that are said to make them invisible to predators below. Their light blends with the dim daylight streaming down, making their body lines indistinct. Other animals that live in deeper, darker waters are thought to use light to attract prey, to signal for a mate, just as fireflies do, or to avoid and confuse predators. Many bacteria floating in seawater are bioluminescent. That seems to help them be eaten, which is just what they want. They take up benign residence in someone's gut or other body part, and both parties benefit. Headlight fish, for example, have sacs of bioluminescent bacteria beneath their eyes. The fish provide the bacteria with a home, and the bacteria, it is thought, shine brightly enough to help the fish shearch for prey.

On the other hand and contrariwise, the bioluminescence of dinoflagellates, those tiny organisms that make the ocean waves glow at night, a sight familiar to anyone who has sailed, is said to be protective *against* being eaten by small crustaceans. Giant squid, which at sixty feet in length — the size of a school bus — are the largest invertebrates alive, can secrete bioluminescent ink clouds that supposedly hide them when they are chased by a predator such as a sperm whale. These purposes and many more are suggested for the enormous numbers of flashings and glowings that go on underwater; one estimate has it that more than 90 percent of living things in the deep ocean are luminescent. But the truth is, say the researchers I talked to, no one really knows or understands the uses of living light very well. Our eyes and our threshold of perception is very different from those of the animals that have the talent, so we don't know exactly what they are seeing, and it is nearly impossible to run controlled experiments in deep water. Still, that doesn't stop our brains from working overtime to make up reasons we find satisfying for a kind of light different from any of our manufacture.

A lot of the sea creatures that have already appeared in these pages — including nudibranchs, hydrozoans, sea cucumbers, and starfish — count bioluminescence among their abilities. Tropical pyrosomes, which are tunicates, live in colonies; stimulation of a single one makes the entire group glow.

Only a couple of freshwater animals are known to be bioluminescent. One freshwater snail knows how to shine, as do some firefly larvae, which noodle about among plant roots in the mud of ponds. This paucity is curious, and no one quite understands it. After all, many freshwater lakes are as dim or even as dark as parts of the ocean.

Although the biochemical pathways vary in different organisms, the production of living light is a straightforward chemical reaction and can be simulated in a test tube. It involves a substrate called luciferin; an enzyme called luciferase; oxygen; and

adenosine triphosphate, or ATP, which is the most important chemical engine we know, involved in every energy transformation in every living thing. The chemical reaction created
with this recipe is an unstable one, and as the molecules seek a
new balance, a photon of light is given off as a byproduct of the
process.

Living light involves such basic chemistry and is so widespread
among completely different animals that some think its origins
may go back to life's beginnings. Howard H. Seliger, who was a
biochemist at Johns Hopkins in the 1960s, is one of the best-
known theorists along those lines.

No one knows exactly what the earth's first atmosphere was
like, but there is good evidence that it contained little oxygen. So
the first life to evolve on this planet, hiding in the ocean water
from the unshielded ultraviolet radiation (there was no ozone layer
then), was anaerobic — it used no oxygen at all. It fed upon the
original chemicals, but when they were gone, some organisms
managed to manufacture their own, using the sun's energy in photosynthesis, as Ralph Lewin explained in verse some pages back.
However, a byproduct of photosynthesis then, as it is today, was
oxygen, which is toxic to anaerobic life. Seliger's idea was that
bioluminescence was just an incidental discharge of energy in a
chemical reaction that detoxified oxygen as it began to accumulate
on the planet. Looked at in that way, it is one of those spandrels
Stephen Jay Gould talks about. Seliger himself revised his hypothesis, and there have been many critics of it over the years. Nevertheless, the idea persists that even if living light had not a single origin but many, it probably came about as an accidental product of a
chemical reaction serving other purposes and has persisted among
an array of unrelated organisms because it later served adaptive
functions.

I telephoned Albert Carlson, a researcher on fireflies at the State
University of New York at Stony Brook, and asked his help in un-

derstanding how living light came about and how it is used by firefly larvae.

"In its initial state," he said, "it is activated by ATP, as you have discovered; it involves a modified amino acid and an activating enzyme. So it may be as basic as protein synthesis. We don't know whether it had a single origin or multiple ones, but whatever the origins it is very widespread, seen among many phyla, which tells us that it is extremely *easy* to evolve.

"Firefly larvae have setae on their cuticle that are pointed backward. I have a student who is working on them, and she has discovered that when the setae are pushed forward, the lantern turns on. Even stimulation of the nerves controlling the setae activates the lantern. If a predator attacks the larva, these setae would be bent, triggering a bright warning glow from the larval lantern. I think that firefly luminescence developed as a protective mechanism in larvae to advertise their unpalatability and was later adapted by the adults for other uses."

"Hmmmmmm," Helen Ghiradella mused when I told her what her colleague had said about adults using larval light for their own purposes. "If that is so, I wonder why the adults grow brandnew lanterns and discard the larval ones." Helen is a professor of biology at another branch of SUNY, at Albany. I visited her at her home near there. Small and trim, with short, dark, curling hair, she has a disarming, iconoclastic, friendly manner. I knew her to be a person with wide-ranging interests. She is also a linguist and an accomplished professional musician; she brings an aesthetic sense to the study of zoology. She is a very *considering* kind of woman.

Puzzling over Al Carlson's observation, Helen explained that the lanterns of firefly larvae give off a slowly rising and slowly fading glow but that the adults signal quickly in bursts of light. She has investigated the off-on switches of that light in adults and has found that they are quite different from those of the larvae. When a larva pupates, it retains its own special lantern but also assembles a new one for adult purposes. For a short time after it emerges from pupation, it keeps the larval lantern along with the adult one, but in a matter of hours, the softly glowing juvenile lantern disappears, leaving only the one capable of sharp flashes.

The larvae I'd seen in Tennessee, in Missouri, in Washington, she told me, probably were young fireflies of the genus *Photuris,* although they might have been genus *Pyractomena.* The only person who could tell me for sure, she added, would be James Lloyd, taxonomist at the University of Florida.

"They are mysterious beasts," she said. "In our area *Photuris* has at least a two-year life cycle, yet larvae cannot be found during the few weeks when adults are active. We don't know whether the larvae are about, but keeping their lanterns dark, or if they go and estivate [the warm-weather equivalent of hibernating] somewhere. The adults also have complicated and incomprehensible — to me — behaviors.

"As to *why* the larvae light up, I think the best explanation is the luminescent equivalent of warning coloration, to tell potential predators not to mess with them. Tom Eisner says they are chemically protected, and I find that dissecting one sure gums up the forceps."

Helen Ghiradella's and Al Carlson's surmises about the function of larval light have very recently been proved by Todd J. Underwood, a Delaware researcher. He and his collaborators found that mice don't like the taste of firefly larvae and won't eat them. What is more, when the researchers soaked Rice Krispies in a bitter substance and offered them up in the company of a light-emitting diode, the mice, having learned to associate light with Bad Dinner, would reject any food when light was the side dish.

Tom Eisner, the Cornell scientist, he of the millipede-hurling mongoose, gave a group of thrushes innocuous mealworms, adult *Photinus* fireflies, and mealworms doused in *Photinus* extract. The thrushes scorned the fireflies and ate the untreated mealworms. If

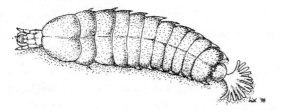

A firefly larva (4 × lifesize).

by accident they did eat a firefly, they threw it up. They also threw up after being tricked into eating the mealworms treated with firefly extract. Analysis showed that the fireflies' chemical protection was in their blood, or hemolymph. Many insects produce defensive substances, but the ones present in the fireflies' hemolymph are chemically unusual. They are steroidal pyrones, which Eisner and his coresearchers named lucibufagins, a word that has buried within it the scientific name for the toad family, Bufonidae. Heretofore the substance had been associated with toad venoms. Never before had such a substance been found in invertebrates. What was the source of lucibufagins? Eisner didn't know; his paper concluded, "Surprisingly little is known about the food of *Photinus*. . . . Dietary origin cannot be ruled out. But it seems improbable that fireflies come upon their defenses by feasting on toads."

It was June when I visited Helen, and there were larval fireflies around Albany. She had captured a few for me and put them into a plastic fast-food container filled with damp earth. The larvae had burrowed into this mudpie, and the place where each one lay was marked by clusters of pinprick-sized breathing holes. Helen also had two other larvae in a plastic cup, which we put under the microscope in her upstairs workroom. With its help I watched them parade around the edge of the cup, turning their heads this way and that, questing in this unlikely plastic environment. Their heads were long, and their backs were covered with what appeared to be a series of overlapping plates, rather armadillo-like, that gave them the scalloped look of a piece of fancywork. They were soberly colored but dappled and spotted, with an overall golden sheen. They were strikingly beautiful. Their bodies ended with a tufted paired pygidium so elaborate and specialized that it is not called simply that, as it is in other invertebrates, but pygypodia, a name that hints at a footlike nature. The pygypodia twisted and turned as the larvae circled; they used them as both prop and propeller while ambling along. Helen handed me a toothpick and told me to turn one of them over with it. The larva's underside was creamy white, and toward the tail end, even in the bright light of the microscope, I could see the two luminescent lantern spots. The larva was writhing about and twisting its pygypodia to try to right itself from its undignified position, but I ruthlessly kept it in place long enough to see, near its head end, behind the antennae, a set of formidable fangs. These larvae are serious predators.

When I left, carrying my fast-food mudpie, Helen also handed me the capped container with the pair of larvae I'd watched under the microscope. The trip back to Washington was too much for one of them, but the other I put into the mudpie and threw in a few twigs and old oak leaves to make the container homier. In a couple of weeks the larvae had to make the trip to Maine with me

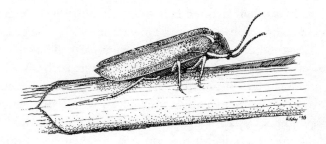

An adult *Photuris* firefly (4 × lifesize)
immediately after emerging from pupation.

in hot weather, which was not good for most of them. After I arrived I discovered that their little air holes had disappeared. But I gently lifted up one of the oak leaves and found that the larva from the plastic cup, the quester, had excavated a saucer-shaped hole and covered itself with leaf bits to pupate. Now ivory in color, it was lying curled on its side with its lantern glowing in the leaf shadow. I could see what looked like the beginnings of wings, mere stubs. On July 3, I found my firefly, now an adult, walking about on the tops of the leaves, so the illustrator was able to draw the picture of it on page 204. It was a fine, new, perfect, fresh-looking adult with black-brown wing covers set off with yellow stripes. Its head was shaped like a flattened yellow shield with a red-brown triangle in its center. It was walking gingerly, as if to get its bearings. By the next day it looked ready for a world larger than take-out. That evening, with the nearby town display of sparklers, Silver Bees, rockets, Hens Laying Eggs, crackers, Earth Thunder, Roman candles, squibs, catherine wheels, and sizzlers noisily lighting up the night sky, I held my own private Fourth of July celebration by releasing the firefly in the grass near the garage. It clung to a blade for a while and then, on new wings, flew up into the air, its bright, cool, silent light flashing its intention to mingle its New York genes with those of Maine. As I watched it, I remembered something Helen had said to me: "I simply can't get used to the wonder of an animal that handles light so easily."

There are very few bugs that people other than invertebrate zoologists are fond of. Perhaps monarch butterflies, ladybugs, and katydids. Out of the suspected tens of millions of insects, those aren't very many. But fireflies make people smile, I've noticed. We all seem to remember chasing them when we were children, capturing a few, and putting them into a glass peanut butter jar with holes punched in its metal lid. Fireflies are the pandas of the insect world. But, as with pandas, their life habits are not particularly cuddly.

Recently a correspondent wrote to the Q & A column on the

gardening page of the *New York Times* and asked what she could plant to attract fireflies. The column's editor turned the question over to the Florida firefly expert James Lloyd, who must have disillusioned the woman by telling her that fireflies are not vegetarians to be attracted to specific plants. He did not pass up the chance, however, to advise, "Let everything go back to nature. What North American fireflies really like best is woodlands, marshes, old fields, and similar, minimally disturbed habitats."

Lloyd had disabused the woman of the notion that she could attract fireflies, as she had butterflies, by planting flowers, but he did not relay the shocking information of what they do eat. He could have, for Lloyd is known within the trade for having applied the term *femme fatale* to certain adult *Photuris* females who flash a species-specific signal to attract mates. However, once mated, they flash in another way, attracting males of the genus *Photinus,* which they then eat. They are exceptional, because most adult fireflies don't spend much time eating. It is the larvae, which need food energy for growing and transforming, who are the champion carnivores. I had been interested in the fangs on the larva I flipped over so rudely under Helen's microscope because I'd been reading about their eating habits and had talked with another entomologist who has studied them.

The females of the European glowworm, *Lampyris noctiluca,* stay wormlike and winglessly larviform during their entire lives although they possess mature reproductive capabilities. That species' eating habits have been studied the most thoroughly, if anecdotally, but evidence suggests that our North American species are similar in their way of feeding.

No one has ever matched the French naturalist Henri Fabre's descriptions of invertebrate behavior, and I am not going to try. Fabre (1823–1915) was so poor that he couldn't afford even the modest equipment that most naturalists of his time had, let alone the expensive piles of stuff with which we equip laboratories today. Somewhere in his writing he reports in a matter-of-fact way

that since he did not have the money for a thermometer, he relied
on temperature readings sent to him by post from a nearby town.
He made up for his material lacks by developing such acute powers of observation that a good deal of our understanding of invertebrates goes back to his descriptions. Here is what he wrote, in
the last year of his life, about a glowworm attacking and de-
vouring a snail.

> The glow-worm for a moment examines the prey, which, accord-
> ing to its habit, is wholly withdrawn into the shell, except the edge
> of the mantle, which projects slightly. Then the hunter's weapon is
> drawn. . . . It consists of two mandibles bent back into a powerful
> hook, very sharp and as thin as a hair. The microscope reveals the
> presence of a slender groove running throughout its length. . . .
> The insect repeatedly taps the snail's mantle with its instrument.
> This is done with such gentleness as to suggest kisses rather than
> bites. When we were children teasing one another, we used to talk
> of "tweaksies" to express a slight squeeze of the fingertips, some-
> thing more like a tickling than a serious pinch. . . . The *Lampyris*
> doles out his tweaks. He distributes them methodically, without
> hurrying, and takes a brief rest after each tap, as if he wishes to as-
> certain the effect produced. Their number is not great; half a dozen
> at most, to subdue the prey and deprive it of all power of move-
> ment. . . . The first few . . . are enough to impart inertia and loss of
> all feeling to the mollusk . . . [by] the *Lampyris,* who . . . instils some
> kind of poison by means of his grooved hooks. Here is proof of the
> efficacy of those sudden twitches, so mild in appearance: I take the
> snail from the *Lampyris,* . . . I prick him with a fine needle. . . .
> There is no quiver of the wounded tissues, no reaction against the
> brutality of the needle. Here is something even more conclusive:
> chance occasionally gives me snails attacked by *Lampyris* while they
> are creeping along, the foot slowly crawling, the tentacles swollen
> to their full extent. A few disordered movements betray a brief ex-
> citement on the part of the mollusk, and then all movement ceases.
> . . . Is the snail really dead? Not at all, for I am free to resuscitate the
> seeming corpse. After two or three days of that singular condition,

which is no longer life and yet is not death, . . . my prisoner . . . is restored to his normal state. . . . . What name shall we give to that form of existence which for a time abolishes the power of movement and the sense of pain? I can see only one that is approximately suitable: anesthesia. . . . Sudden and profound anesthesia is an excellent means of enabling the *Lampyris* to attain his object, which is to consume his prey in perfect quiet. What is his manner of consuming it? . . . The glow-worm does not eat in the strict sense of the word: he drinks his fill; he feeds on a thin gruel into which he transforms his prey. . . . He is able to digest before consuming; he liquefies his prey before feeding on it. . . . By repeated tiny bites, similar to the tweaks which we saw distributed at the outset, the flesh of the mollusk is converted into a gruel . . . the consuming of which will occupy him for several days. . . . When he leaves the table, the shell is found to be completely empty.

Wondering if anyone had followed up on Fabre's observation, I found that Jonathan Copeland, a biologist working at Georgia Southern University in Statesboro, had done some research on the effect of extract of firefly larvae on snails. He found that it stopped the heartbeat but that it was reversible, just as Fabre had reported. I asked him if the terrible stuff that in small doses anesthetized a snail and in larger ones turned it into soup was the same lucibufagins Tom Eisner had written about and if it also protected larval fireflies from predators. "Well, we don't know," Jon replied, "because no one has done any more work on it, but it is in the hemolymph, and I suspect it is the same. It certainly is an attractive hypothesis."

Al Carlson thought so, too, and went on to say, "It is possible they can get it from food: that's common enough among beetles. I have a colleague who is working on beetles that feed on deadly nightshade and have an everted anus which allows them to spread poisonous fecal pellets on their backs as a result. But firefly larvae could modify an innocuous substance and produce something with poisonous qualities if they can get the basic molecules. They have that ability. There has been no work done on their food, so we'll have to wait until someone does.

Months after my Firefly Fourth, Tom Eisner and colleagues published a new paper reporting that *Photuris* fireflies in their adult form possess no lucibufagins at all, but that the females (and possibly the males) acquire them by eating *Photinus,* which do possess them. Female *Photuris* fireflies were the ones Jim Lloyd had dubbed *femmes fatales.* Apparently they weren't deceptively attracting and eating the *Photinus* males just for nutrition, as had once been thought, but for the protection their chemical makeup afforded. What is more, Tom told me, after I telephoned him at Cornell to ask about the research, the female *Photuris* could pass along the acquired lucibufagins to their eggs. What about the larvae that develop from them? Did they have lucibufagins, too? Might that be the substance that gave *Photuris* larvae their formidable anesthetizing and predigestive abilities, which arrested cardiac function in snails?

"No," said Tom firmly, "it must be something else, because the *Photuris* larvae have no lucibufagins in their bodies at all."

The mystery remains.

And however it is that a firefly larva acquires its chemical makeup and the pyrotechnic ability that advertises it, I would not like to be a snail (or an earthworm or a slug — reports say the larvae attack them too) in a firefly world. As G. K. Chesterton wrote in 1901, "A beetle may or may not be inferior to a man — the matter awaits demonstration; but if he were inferior to man by 10,000 fathoms, the fact remains that there is probably a beetle view of things of which a man is entirely ignorant."

READINGS

Alberte, Randall S. "Bioluminescence." *Naval Research Reviews* 45 (1993).
Blum, S. Murray, and Arumagin Sannasi. "Reflex Bleeding in the Lampyrid *Photinus pyralis." Journal of Insect Physiology* 20 (1973).

Buck, John. "The Anatomy and Physiology of the Light Organ in Fireflies." *Annals of the New York Academy of Science* 49, art. 3 (1948).

Campbell, A. K., et al., eds. *Bioluminescence and Chemoluminescence.* Proceedings of the eighth international symposium. New York: John Wiley & Sons, 1994.

Carlson, Albert. "Factors Affecting Firefly Larval Luminescence." *Biological Bulletin* 129, no. 2 (1965).

————, and Jonathan Copeland. "Communication in Insects. 1. Flash Communication in Fireflies." *Quarterly Review of Biology* 60 (December 1985).

————. "Behavioral Plasticity in the Flash Communication Systems of Fireflies." *American Scientist* 66 (1978).

Copeland, Jonathan. "Effects of Larval Firefly Extracts on Molluscan Cardiac Activity." *Experientia* 37 (1981).

————. "Neuroethology of Prey Capture by Larval Fireflies." *Society for Neuroscience Abstracts* 6 (1980).

Crowson, R. A. *The Biology of Coleoptera.* New York: Academic Press, 1981.

Domagala, Peter, and Helen Ghiradella. "Structure and Function of the Terminal Appendages of Photurid Firefly Larvae." *Biological Bulletin* 166 (1984).

Eisner, Thomas, et al. "Firefly 'Femmes Fatales' Acquire Defensive Steroids (Lucibufagins) from Their Firefly Prey." *National Academy of Science Proceedings* 94 (September 1997).

Eisner, Thomas, et al. "Lucibufagins." *National Academy of Science Proceedings* 75 no. 2 (1978).

Fabre, Henri. "The Glowworm." *Century Magazine* 87 (1913).

Ghiradella, Helen. "Anatomy of Light Production: Fine Structure of the Firefly Lantern." In *Insects,* ed. Michael Locke, vol. 11 of *Microscopic Anatomy of Invertebrates,* ed. Frederick W. Harrison. New York: Wiley-Liss, 1998.

————. "Fine Structure of the Tracheoles of the Lantern of a Photurid Firefly." *Journal of Morphology* 153 (1977).

Goetz, M. A., et al. "Lucibufagins, IV." *Experientia* 37 (1981).

Harvey, Edmund Newton. 1940. *Living Light.* Princeton, N.J.: Princeton University Press, 1940.

Hastings, J. W. 1983. "Biological Diversity, Chemical Mechanisms, and the Evolutionary Origins of Bioluminescent Systems." *Journal of Molecular Evolution* 19 (1983).

Herring, Peter. "Systematic Distribution of Bioluminescence in Living Organisms." *Journal of Bioluminescence and Chemoluminescence* 1 (1987).

————, ed. *Light and Life in the Sea.* Cambridge, Eng.: Cambridge University Press, 1990.

————. *Bioluminescence in Action.* New York: Academic Press, 1978.

Hess, Walter Norton. *Studies on the Lampyridae.* Ithaca, N.Y.: Cornell University Press, 1922.

Lloyd, James E. "Bioluminescence and Communication in Insects." *Annual Review of Entomology* (1983).

McElroy, W. D., and H. H. Seliger. "Origin and Evolution of Bioluminescence." In *Horizons in Biochemistry,* ed. Michael Kasha and Bernard Pullman. New York: Academic Press, 1962.

McLean, Miriam, et al. "Culture and Larval Behavior of Photurid Fireflies." *American Midland Naturalist* 87, no. 1 (1972).

Morin, James. "Coastal Bioluminescence." *Bulletin of Marine Science* 33, no. 4 (1983).

Newport, George. 1857. "Natural History of the Glowworm." *Linnaean Society, Zoology, Proceedings Journal* 1 (1857).

Phipson, Thomas L. *Phosphorescence.* London: Lovell, 1862.

Rose, Kenneth J. "Living Lights." *Science Digest* 92 (1984).

Schopf, J., et al., eds. *Proterozoic Biosphere.* Cambridge, Eng.: Cambridge University Press, 1992.

Seliger, H. H. "Bioluminescence." *Naval Research Review* 45 (1993).

———. "Evolution of Bioluminescence in Bacteria." *Photochemistry and Photobiology* 45, no. 2 (1987).

———. "Origin of Bioluminescence." *Photochemistry and Photobiology* 2 (1975).

"Smoking Makes Your Blood Glow." *Science News* 124 (Oct. 29, 1983).

Underwood, Todd J., et al. "Bioluminescence in Firefly Larvae: A Test of the Aposematic Hypothesis." *Journal of Insect Behavior* 10 (1997).

Williams, Francis X. "Notes on the Life-History of Some North American Lampyridae." *Journal of the New York Entomological Society* 25 (1917).

Young, Richard E. "Oceanic Bioluminescence." *Bulletin of Marine Science* 33, no. 4 (1983).

13

THE BICYCLE I'D BOUGHT when I first moved to Maine became a comfortable companion over time. It regularly took me exploring down back roads and on errands into town. Before long it had a partner, and when Arne came up from Washington on fair-weather weekends, we bicycled together.

One weekday morning after the summer people had left, and after I had been egging with Everett West, I wheeled my bicycle out and let it take me to the park by the ocean. The weather had been softly foggy and rainy for several days, but the morning had brought one of those high-pressure days in which the sky is blue, clear, and clean. The low autumnal angle of the sun made the world seem both brand-new and serenely ancient at the same time. The light had pulled me from my desk, made it impossible not to be out of doors. The weather was cool enough to put on a windbreaker and gloves. The fragrance of wet duff, live spruce, and fir mingled with the scent of ocean as I pedaled along.

The tide was out, as it had been the first time I visited the park, and I clambered down the granite cliffs, followed the black rucked-up basalt strip to the cobble, and poked around the edges of the biggest tide pool. It was more familiar to me than when I had first come, for I had visited it often and learned more about its

moods and its inhabitants. But, as is always the case, small knowledge had brought me only to the we-don't-knows and the no-
work-has-been-done-on-it of researchers, and the sense of mystery and honor had grown. When I picked up a curly piece of
cream-colored algae, a little amphipod, looking like a miniature
shrimp, leaped from it, bounding from one of my fingers to another. I knew its name — or, more precisely, the name we humans have given it — *Hyale nilssoni,* for one day months earlier I
had spent some time identifying it. It is an isopod, as are the pill
bugs and relatives of the sand fleas that hop about busily feasting on
beached seaweed. It has a one-year life span, I'd read, but I don't
know what it packs into that year, so having a name for it hardly
signifies. If I've gained some few facts since my first visit to the tide
pool, they have simply enlarged the scope of my ignorance. I held
my hand over the water and encouraged the amphipod to hop
back in among the barnacles, where it prefers to live.

When I'd first started coming to the park, I'd hoped I would
find a sea mouse, *Aphrodite aculeata,* an iridescent furry worm that
grows to six inches in length. I knew that they live in the cold Atlantic waters, though they were not on the census list of animals
found in the park's critical area. When I talked to Peter Larsen,
who had taken the census, he said, "Oh, yeah. They're out there,
but there isn't much left of them when we bring them in." I'd
given up expecting to find them at the park, for I'd learned that
they live in the muddy bottom of deeper ocean waters, and it
would be highly unlikely that they'd wash into a tide pool.

Actually, by the time of my autumn visit to the park, I'd seen
specimens of several species of *Aphrodite* floating in preserving
fluid in big jars at the Smithsonian's Natural History Museum.
Kristian Fauchald, chairman of the Department of Invertebrates
and a specialist in polychaete worms, of which the Aphroditidae
are one family, had shown them to me one afternoon months before. Oval and muscular looking, they were covered with dense

furry coats, except for their bellies, which were naked and clearly marked with lines of segmentation. When alive they had pumped fresh, aereated water over that naked underside. The coats on some specimens were matted to a feltlike consistency and caked with mud; some wore cleaner, more velvety garb. I asked Kristian if they were soft to the touch. He said they were; I could also see protective spines poking out from the felt. Kristian handed me a jar containing *A. aculeata*. Its coat was golden around the edges and shimmered with blue and green iridescence as I turned the jar this way and that in a sunbeam coming through the window. It was a beautiful animal.

On that bright autumn day in Maine, I left the tide pool and clambered up through the cobble and small rock, found myself two boulders that made a granite chair, and sat down. I peeled off my windbreaker in the light breeze and warming sun. Buoys marked the lobster traps in the water. The purr of a fisherman's boat engine trailed behind him as he stopped to check the traps on his way down coast. After he rounded the point, I could hear nothing but nearby gull conversations and the lap of gentle waves as the tide continued its retreat. I was facing east, and the sun's shine made a dappled silver path leading from the rocks below me out between two islands to the ocean horizon, taking my eye and my thoughts with it.

One of our neighbors, a lobsterman in his eighties, had paid a social call with his wife the evening before, and I'd asked him if he ever found sea mice in his traps. He smiled in recollection and said that yes, golden-brown they were, and kind of shiny-like. He usually found one or two in the course of a twelve-month; hadn't seen any this year so far. But if he did, he'd bring one to me.

No one forgets a sea mouse after he's seen it. They are animals much noticed but little known. In reading up on them, I'd found that even the scientific descriptions grew lyrical. "Reddish bronzy . . . brilliant bluish green," wrote Marian H. Pettibone, the grand old lady of polychaete worm studies in a technical paper describing specimens taken near my park. A nineteenth-century British authority, William McIntosh, let the word "gorgeous" slip into his scientific description of the worm and lamented that "it pursues its work in the depths of the sea where its beautiful iridescent hairs can be seen by few admirers." Another British marine zoologist of the same century, Philip Henry Gosse, even reached for simile: "The Sea-Mouse . . . is clothed with a dense coat of long bristles, which are fully as resplendent as the plumage of a Humming-bird."

The iridescence, Kristian told me, is purely mechanical, the result of the positioning of strands of collagen in the hairs. Younger specimens looked duller because the strands were spaced differently. When I asked him about reports I'd read that they were bioluminescent, he replied, "I've never seen it myself, but I believe they are. It may be borrowed, however, because they are hosts to ostracods [small crustaceans sometimes called sea shrimp] or bacteria that are bioluminescent."

Before my meeting with Kristian, I had read what I'd been able to find on the Aphroditidae, including a monograph by Mahalah Glen Clark Fordham, published in 1925, which Kristian had recommended as being just about the only sound work on them. "Unfortunately," he had written to me, "not much has been published about these worms that would make a lot of sense."

Kristian is one of the most highly regarded polychaete specialists in the world, but to thousands of zoology students he is known for his observation, at the head of the chapter on Annelida, the worms, in a standard textbook: "The study of polychaetes used to be a leisurely occupation, practiced calmly and slowly." Kristian wears his authority lightly. He is a man with a friendly face that has seen weather. His sandy hair and beard are graying. There is a hint of his native Norway in his speech.

"Ignorance!" he said cheerily as he leaned back in his chair with his hands behind his head. "Confusion!!!" he added, laughing. I had shown him a photocopy of the earliest known extant Western entry on an aphroditid, published in 1554, in which the French physician Guillaume Rondelet named it physsalus, taking his cue from a surgeon of Roman times. The name is from the Greek word — *physa* — for a blast of air, because Rondelet believed, as had the Roman, that the animal could swell with air at will. Rondelet, the reader may remember, was also the man who gave us the confusing term "Aristotle's lantern."

When Kristian laughs he dimples, and he dimpled at me and said, "You've got to understand that no one has ever seen *Aphrodite* except when it was unhappy." The animal Rondolet was describing, he said, was one that had died, putrefied, and bloated. He skimmed the rest of the Latin text, chuckling to himself, especially over the accompanying woodcut, which bore little resemblance to the lovely animals I'd seen in the jars. "Well, maybe a sea mouse upside down," he said squinting at it experimentally with one eye.

A complete translation of the 1554 Latin text is worth giving because it reveals the muddle in which the animal entered the scientific literature. The translation was made by John Campbell, a classicist from the University of South Florida. A copy of the original woodcut, with a condensed entry in French from the 1558 edition, is on page 217.

Concerning the Physsalus

Discussing the fish of the Red Sea, Aelianus recounts Leonides' observations on the physsalus, which he describes in the following words. Leonides the Byzantine writes that the fish is native to the Red Sea and the Persian Gulf, that, no smaller than a mature gudgeon, without eyes and mouth, it is nevertheless endowed with fins like a fish; that the semblance of a head, although less clear, can be divined; and that a certain shape beneath the stomach appears to be drawn gradually into a bosom, which has the color of an emer-

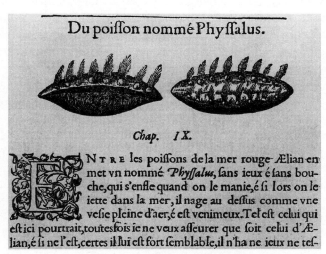

Du poiſſon nommé Phyſſalus.

Chap. IX.

NTRE les poiſſons de la mer rouge Ælian en met vn nommé *Phyſſalus*, ſans ieux é ſans bouche, qui s'enfle quand on le manie, é ſi lors on le iette dans la mer, il nage au deſſus comme vne veſie pleine d'aer, é eſt venimeux. Tel eſt celui qui eſt ici pourtrait, toutesfois ie ne veux aſſeurer que ſoit celui d'Ælian, é ſi ne l'eſt, certes il lui eſt fort ſemblable, il n'ha ne ieux ne teſ-

The woodcut illustration of *Physsalus* from the 1558 edition of Rondelet's *L'Histoire entière des Poissons*.

ald, which he wrote serves for eyes and a mouth. Whoever swallows one, discovers that he has caught a fish that promises dire consequences; for, as soon as he has tasted it, he suffers sharp pains; then its stomach bursts, producing a serious threat to life; in any case, if ever caught, it exacts a terrible price. As soon as it is out of water, it inflates, and, if you touch it, swelling more and more, grows hot. And then if anyone continually poke it, the entire creature grows translucent because of the agitation, and finally ruptures as if afflicted by dropsy. If someone should wish to return it still alive to the sea, it swims on the surface like a bladder filled with air. For this reason Leonides writes sentimentally that it is called the physsalus. He also places in the same family the physa fish, which nevertheless he distinguishes from the physsalus, lest anyone think they are the same because of the similarity of their names; for the physa is an Egyptian fish endowed with an extraordinary nature; indeed they say it possesses awareness: when the moon waxes and wanes, its liver swells and subsides along with the moon, while the body grows first fat and then thin and slight. We do not assert that the fish described here is exactly the same as the physsalus, although some fish, believed to be native to certain places, are nevertheless found occasionally, however rarely, in other places. But if it is not the physsalus, it can be justly said that it is certainly not very different from it; or, if you prefer, it can be more properly placed in the family of sea worms. It lacks a mouth and eyes, is wider in the middle, grows thin in the extremes where it curves, is full of wrinkles on the stomach or underside, with the suggestion of female genitals; on the sloping side or on the back, small protuberances project, which our fishermen call warts, from which green hairs emanate. But when touched it swells and swims on the surface, as Aelianus wrote concerning the physsalus. Moreover we have discovered that it is poisonous in the dog days of summer.

Rondelet's misinformation has been repeated down through the centuries to this very day. Later authors, even though they discarded the name physsalus, still maintained that the animal could swell up. And many repeated the statement that it was poisonous. A modern encyclopedia of marine invertebrates for sale in the Smithsonian's own bookshop described it as having "poisonous spines."

"NO!!!" said Kristian. "Not poisonous. *Aphrodite* has no chemical protection at all. Perhaps it got mixed up with fire worms, *Hermodice carunculata,* which have setae that are harmful to touch."

In the ensuing centuries, more accurate observers such as Fordham, Pettibone, and a few others, have measured and dissected various species of *Aphrodite* and described their anatomy in detail. But what the animal does with that anatomy remains largely a mystery. "*Aphrodite* is not a shore form," Fordham wrote, "and it rarely lives in confinement for more than five or six days."

"They are difficult to find in the ocean," Kristian added, "because of the depth at which they live, and all mud dwellers are hard to raise in captivity because the tanks you put them in become anoxic quickly and in a nasty way."

What little is known about them can be stated briefly. Sea mice are carnivorous worms made up of forty to fifty segments. They search the soft mud in cold ocean deeps for their food. They take it up in a manner not uncommon among other invertebrates but one that seems unusual to us: they send their gut out for dinner. They possess a strong, muscular pharynx, which they are able to evert from their bodies. With it, by force of four surrounding jaws, they crush whatever they can find. They are not "fierce and rapacious, overpowering creatures incapable of resistance," as Sir John Dalyell, writing in the mid-1800s, said in his curious guide to the "Humbler Tribes of Animated Nature."

"They suck and gum their food," Kristian said. He explained that they take in a lot of mud and that the nutritious parts are quickly absorbed while the mud is more slowly processed and excreted.

Sea mice belong to the group of polychaetes called scale worms, and hidden under their furry coats they have scalelike flaps called elytra, with which they can paddle and pump to keep fluids flowing within their bodies. The pumping action, which involves the whole body, is rhythmic and graceful, as I was able to see for myself eventually. At the end of the body is the anus, which is opened on the downstroke to allow fluid to pass through and is closed on the upstroke; this rhythmic pumping keeps fresh, oxygenated water flowing around the animal and over the bare spots on its thin-walled body, which absorbs oxygen readily. Oxygen levels in the muddy bottoms of the ocean deeps are low, so the ability to pump fresh water through themselves is important. In addition, Kristian told me, they need less oxygen than we do; they can live at oxygen levels we would find intolerable. Their felt covering, he said, helps keep their body pores free of clogging debris, but "it also keeps liquids from leaking out of pores in the wrong places. It does form a real felt, and if you've ever walked on felt soles you know how waterproof it can be."

In the mud, the sea mouse burrows and creeps along on bifurcated appendages called parapodia, from which sprout spines and those tufts of golden hairs. The parapodia are used alternately, right, left, right, left, right, left, giving the animal's progression a graceful, sinuous quality. Despite Rondelet's assertion to the contrary, sea mice do have eyes, and their heads also have a variety of what seem to be sensory organs, including antennae and a lumpish "facial tubercle," the workings of which are completely unknown.

Unlike their distant cousins the hermaphrodite earthworms, sea mice come in two genders, and eggs and sperm are produced from the sex cells lining the blood vessels. When ripe, the eggs and sperm float out through pores called nephridia, simple canals leading from inside their bodies to the outside. No one knows what happens next. Fordham had ventured that the females may hold and brood the fertilized eggs under their elytra, and Kristian said that was a good guess. "The eggs contain a lot of yolk," he said, "and that is characteristic of an egg that is brooded, but we really don't know."

And what happens as the egg develops is also unknown. Does it hatch out as a larva, or trochophore, in zoological terms? No one knows. Does it hatch into a small but recognizable sea mouse? Young forms have been seen, but not egg-sized ones. No one knows.

In fact, so little is known about their reproduction and youthful development that sea mice might just as well be born of foam, as their eponym was. It was Linnaeus who gave this lovely golden worm the name we use for it today, *Aphrodite aculeata,* spiny aphrodite, although in his first edition he called it *A. nitens,* shining aphrodite. Its common name, time out of mind, had been sea mouse or, more fittingly, sea mole. Rondelet's name *Physsalus* did not catch on; scientific folk before Linnaeus's time usually called it *Vermis aurens,* golden worm, or they simply translated the common name into Latin and called it *Mus marinus.* For a time it even carried a name about as long as the animal itself: *Eruca marina Rondeletti pilis in dorso instar colli columbini variegatis,* which is easy enough to understand with small Latin, once you know that *Eruca* is the word for caterpillar. When I began talking with coastal Maine fishermen about sea mice and how much I wanted to see a live one, I soon discovered that this long Latin name still has, in shortened form, currency here. A number of them call aphroditids "sea caterpillars" and reserve "sea mouse" for the long-tailed pink tunicate *Boltenia ovifera,* known commonly in other places as the stalked sea squirt. Such is the problem with common names, but I rather like the fact that some Mainers know *Aphrodite* as sea caterpillar.

I'd assumed that Linnaeus had named it *Aphrodite* because the golden-haired goddess of that name was held to be so beautiful, but I'd overestimated Linnaeus, the specialist in tasteless joke names. Kristian, the Norwegian, told me that Linnaeus, the Swede, was familiar with the centuries-old northern European slang use of the word "mouse" for a woman's sexual parts and that the golden worm, the sea caterpillar, with its "suggestion of female genitals," as Rondelet put it, was commonly called sea mouse for that reason. Linnaeus perpetuated the indignity by naming it for the goddess associated with the physicality of love and creation. It is hard to work up any righteous feminist irritation over this minor obscenity because the last laugh belongs to the worm and, perhaps, to the goddess.

Linnaeus, with his only-nine-thousand-species-of-everything approach to the living world, gathered up all the little slithery creatures that weren't insects and called them worms. He included in this category many animals that we wouldn't call worms today, such as all the shelly creatures from clams to conchs that we find along the shore, as well as sponges, sand dollars, starfish, and a hodgepodge of "extremely minute animalcules, destitute of feelers, and generally not visible to the naked eye." By culling insects from his worm category he was actually being very up to date and scientific, because the word "worm," an ancient one, had been used to mean almost any animal that was not human or of particular regard to humans. As a category, it had encompassed all bugs and beasties, real and unreal, snakes, spiders, sea monsters, and dragons. Beowulf, it will be remembered, saved his people from "the warfare of the great Worm . . . this Ravager . . . this fell Destroyer . . . hated and hounded."

Today we think we've learned better; we've ceased believing in great Worms. We call clams bivalve mollusks, and animals not visible to the naked eye microorganisms, not animalcules. We've tamed the word "worm," and it has become a shrunken thing with the disappearance of dragons. I find from a completely unscientific survey of acquaintances that the word today conjures up nothing but earthworms or those things we medicate our dogs and cats against.

For zoologists the word is richer. The Annelida, the worms, includes leeches, oligochaetes, such as the earthworms that Sam James studies, and polychaetes, Kristian's field. Of the three, the polychaetes are the most numerous, represented by more than ten thousand species, with new ones being discovered all the time. If earthworms are called oligochaetes because they have few chaetae, or setae, or hairy bristles, then Kristian's kind are polychaetes because they have many chaetae or setae. They are woolly, feathery, tentacle-y, dustmoppy animals with many different ways of life. They encompass such surprising specialty and diversity that the word "worm" when used by an annelid expert conjures up almost as many forms as it did one thousand years ago. Many are strikingly beautiful animals, and they come in a rainbow of colors and a large array of shapes, some of which are sketched on page 223. Some stay placidly in burrows in the mud and wave their feather dusters about to catch food; some are graceful swimmers. Some are predatory; some are filter feeders; some feed on plants. Some are chemi-

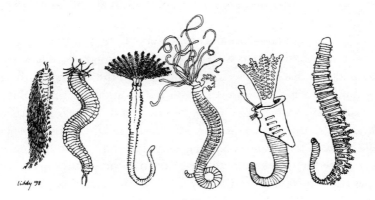

Polychaetes of various kinds. *Aphrodite* is the first on the left.

cally protected; some are not. Their commonality comes from the fact that they are all seagoing annelids, but more subtly it has to do with their growth and evolution, the origins of segmentation, the basic plan for animal development around a hollow cavity, and the formation of that cavity in early divisions of the fertilized egg.

This is very deep stuff, having to do with the beginnings on the planet of metazoans, or multicellular life forms. Sponges represent one kind of metazoan organization, one that didn't lead to anything but more sponges, interesting in its uniqueness. But the plan of building a body around a hollow cavity and then repeating the arrangement in segments became one of life's favorites, repeated with variations over and over again throughout the animal kingdom. For a number of cladistic reasons, zoologists have reasoned that those first sinuous, winding fossil tracks in the Precambrian ocean muds were made by an animal that was polychaete-like, ancestral to the worms that followed and beings of that metazoan structure. "The proposed ancestral polychaete," Kristian wrote in "Polychaete Phylogeny" in 1974, "was a burrowing organism with setae, but without parapodia, adapted to burrowing in soft muds . . . one of the richest environments in the sea in terms of available energy."

My mind does not have the proper setup to find very satisfying a creation myth that starts with the Word. But I might be able to get behind one that has as its topic sentence "In the Beginning there was the Worm." At least, that was my thought on that golden autumn day, with maples red-leafed here and there, as I pedaled home from my visit to the park. And it seemed fitting that *Aphrodite aculeata,* an advanced and specialized polychaete, should echo the name the Greeks gave to the generative and creative principle that had emerged from the sea's depths. Aphrodite, the name, derives from the Greek word for seafoam, *aphros.*

Her story, from Greek sources, begins once upon a time, a time long ago, before there were gods. Everything was mixed up together then. Earth and Sky were one. Sky, Ouranos, was male, and he impregnated Earth, whose name was Gaia, and she gave birth to the Titans, giants, one of whom was Kronos, or Time. Gaia gave Kronos a flint sickle, and when he wielded it the earth and sky were parted. But with that same pass of the sickle he emasculated his father, Ouranos. Drops of blood fell to the earth and from them were born the Furies. Hesiod tells what happened next.

> . . . the members themselves, when Kronos
> had lopped them with the flint,
> he threw from the mainland
> into the great wash of the sea water
> and they drifted a great while
> on the open sea, and there spread
> a circle of white foam
> from the immortal flesh, and in it
> grew a girl, whose course first took her
> to holy Kythera,
> and from there she afterward made her way

to sea-washed Cyprus
and stepped ashore, a modest lovely goddess,
and about her
light and slender feet the grass grew,
and the gods call her
Aphrodite, and men do too
because from the seafoam
she grew . . .

Golden-haired Aphrodite was known to the East as Ashtaroth, in Africa as Isis. She was regarded as the creative, organizing principle, rising from Chaos, dancing upon the sea. The floor of her sanctuaries had to be made of seashells, and the sea urchin was sacred to her. At Athens she was known as the eldest of the Fates, and wherever she stepped, not only grass but flowers grew, and fruit ripened. Later legends ascribe many illicit love affairs to her and emphasize her physicality, her sensuousness, her lust, her amorality. She becomes both common and enduring, a blowsy, fecund Mother Nature. Aphrodite, however, always kept her tie to the sea and periodically renewed her virginity by returning to it, bathing in it, and emerging afresh.

There is one curiously non-Greek part of her story that makes using her name for our hairy sea mouse particularly apt, and that is her aspect as Aphroditos, noted by Aristophanes, Macrobius, Servius, and others. On Cyprus, which, as Hesiod noted, was one of her earlier stopping places, an androgynous, hairy, bearded Aphrodite, known as Aphroditos, was worshipped, a deity with the body and clothes of a woman but with the beard and sexual organs of a man. The general opinion is that this hairy Aphrodite originated to the east of Greece and embodied the concept of a unified yin and yang. It may have been tamed and westernized in the later story about Aphrodite's offspring, Hermaphroditus, a double-sexed being, whose father was Hermes, messenger of the gods. Aphroditos, however, seems to have been a very ancient, pre-Classical expression of the androgynous qualities of an undifferentiated creative principle, of bountifulness, made concrete as a Great Goddess with masculine traits.

Not bad for a six-inch fuzzy iridescent worm of ancient lineage from the ocean deeps.

My neighbor didn't find me a sea mouse, but another fisherman, who'd heard I was looking for one, did. He'd been dragging for sea cucumbers and had hauled up three sea mice while working the muddy bottom of a nearby point at a depth of about sixty feet. "It's funny," he said when he handed them to me in a pail of salt water. "I hardly ever see these, and today I found three." I put them into a gallon jar of seawater and took them to the illustrator so she could make the drawing below. The three were lively, pumping water through their bodies, which curved as they did so, and spurting water from their anuses. They tested the sides of the jar with their parapodia, extending and retracting them in waves of motion.

The next day they weren't so lively. The water in their jar was probably too low in oxygen, and they may have suffered from the change in pressure. So, with a friend, I went down to the park to

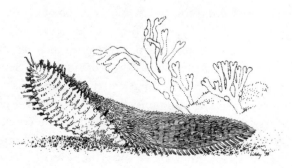

A sea mouse, *Aphrodite aculeata* (½ lifesize).

release them into deep water. I took along a specimen jar and a 70 percent alcohol solution, thinking I might sacrifice one and keep it. I took them out of the big jar, and my friend ran an appreciative finger along their silky fur. He admired the golden shimmer around their edges, and I turned them from side to side to catch the sunlight so he could see their blue-green iridescence. I showed him their antennae, the tubercles on their heads. On the palm of my hand I could feel a tickle as their parapodia extended and contracted. The smallest one was more sluggish than the other two, probably moribund, a good candidate for the specimen jar. But one by one I dropped them back into the ocean. They weren't mine to keep.

READINGS

Beowulf. Trans. William E. Leonard. New York: Heritage Press, 1923.

Dalyell, John Graham. *Powers of the Creator Displayed in the Creation or Observations of the Humbler Tribes of Animated Nature with Practical Comments & Illustrations.* London: Van Voorst, 1853.

Darboux, Jean Gaston. *Recherches sur les Aphroditiens.* Montpellier: C. Coulet, 1899.

Delcourt, Marine. *Hermaphrodite.* London: Studio Books, 1961.

Fauchald, Kristian. "Polychaete Phylogeny, 1974." *Systematic Zoology* 23 (1974; revision in progress).

———. "Diet of Worms." In *Oceanography and Marine Biology: Annual Review* (1979).

Fordham, Mahalah Glen Clark. *Aphrodite aculeata.* Liverpool, Eng.: University Press of Liverpool, 1925.

Gosse, Philip Henry. *Manual of Marine Zoology for the British Isles.* London: Van Vorst, 1885–86.

Graves, Robert. *The Greek Myths.* New York: Braziller, 1959.

Hastings, James. *Encyclopedia of Religion and Ethics.* Edinburgh: Clark, n.d.

Hesiod. *Theogeny.* Trans. Richard Lattimore. Ann Arbor: University of Michigan Press, 1959.

Hutchings, Pat, and Jane McRae. "The Aphroditidae from Australia, Together with a Redescription of the Aphroditidae Collected during the Siboga Expedition (1916 a,b; 1917)." *Records of the Australian Museum* 45 (1993).

Leach, Marjorie. *Guide to the Gods.* Santa Barbara, Calif: ABC–CLIO, 1992.

McIntosh, William C. *Monograph of the British Annelids.* London: Ray Society, 1873–1900.

————. "Report on the Annelida Polychaeta Collected by the H.M.S. *Challenger* during the Years 1873–76." In *Voyages of the H.M.S. Challenger,* vol. 12. Edinburgh, n.d.

Pettibone, Marian H. *Marine Polychaete Worms of the New England Region.* Washington, D.C.: Smithsonian Institution Press, 1963.

Rondelet, Guillaume. *L'histoire entière des Poissons.* Lyons: Bonhomme, 1558.

————. *Libri de Piscibus Marinus.* Lyons: Bonhomme, 1554–55.

Wissowa, Georg, ed. *Pauly's Real Encyclopädie.* Stuttgart: Metzlerscher Verlag, 1894.

Epilogue

Over the course of my time in the Ozarks I made a lot of friends, but there were two families with whom I was particularly close. When I left, both gave me presents. The husband and father of one of those families dug me a fine rock. The spot he'd taken it from meant something to all of us, so it was an appropriate parting gift. But it was also an excellent rock in size, configuration, and texture. The Ozarks are full of rocks just crying out to be built with, and over the course of my life there I did just that with many of them. It became something of a specialty. Once you start working with rock, you look at the world differently and see potential building supply everywhere. Fortunately Maine has as much rock as southern Missouri, and the friend who gave me the Ozark rock knew I would put it with companions. I did. It was one of the first I laid in the brick, rock, and rubble walkway I made from one of the new porches to the driveway; I walk over it every day and remember my friends.

The other family gave me a little sack of acorns, good Ozark red-oak acorns. Those I kept close by me, and a couple of months after that going-away party in Missouri I came up to Maine to spend a month. The builder had finished all the demolition, the new concrete had been poured, and the new house parts were up and enclosed. He and his crew had another job to complete quickly, so they were taking a month off from my work. I came up from Washington because I couldn't stay away.

It was winter, but winter along the Maine coast is milder than it is in the interior of the state, and certainly milder than in the Ozarks, so I could still plant my acorns. I camped out in the deconstructed house, a place of studs and sills, very little insulation, and many cracks, providing a healthy source of fresh air. I'd bought a wood stove; I had a back-up oil burner. I stapled plastic bags over many of the cracks, but it was pretty drafty and brisk. Nevertheless, I could climb up the ghost of a former attic stairway, thread my way out over two beams, and find myself in my new tower. There was sawdust on the partial floor and leftover nails scattered on it, but I propped up some boards and had my morning coffee there, watched the sun rise over the ocean.

The builder shooed me out at the end of the month, and I went back to Arne and Washington when the work crew returned. By June they were finished. There is now a proper stairway up to my tower, and I have a comfortable chair there to watch the beginning of the new day.

When I left the Ozarks I left my chain saw behind, gave it to a friend. I'd rather thought that a woman in her mid-sixties ought to buy, not cut, firewood. But the first summer I was here, while the house was still being taken apart, I realized that the ocean view needed improving and that the woods badly needed rearranging. Besides, a windstorm had swept through the year before and had toppled a number of aging spruce and fir trees, which needed to be tidied. I bought a new chain saw, a smaller one, more suitable to my years. On afternoons when I wasn't whacking out bathroom tile I began thinning the woods and creating paths, making space for some trees, taking out poor ones, producing considerable firewood in the process.

There's a good five years of work with a chain saw here before I even establish the outlines of what I want to do with the woodlot, and time, of course, brings new trees to thin, dying ones to take out, more windstorms. I wonder when I *will* be too old to use a chain saw or if I'll live long enough to create the woods I can see in my head. Already, making firewood from a tree I've dropped, hauling it out, stacking it, and heaping the leftover brush into piles for small-animal condos, is harder work than it used to be. I need more rest breaks.

Certain body parts are wearing out. My own personal time is winding down. In my family we tend to wink out from cardiovascular "events." That's a good way to go compared to some of the wasting disorders that the medico-industrial complex now tries to save us for. We never know the manner or time of our deaths, but my genes are turning on right on schedule.

This *memento mori* makes me value present time ever more, turns it into a shimmering thing of great value, something I spend more carefully than I did formerly. In the 1930s and 1940s I went to a school that was considered progressive. We were not graded by numbers or letters but by categories assured to create a load of self-doubt and guilt, such as "Works to Capacity." "Uses Time to Good Advantage" was one in which my third-grade teacher found me particularly wanting. I'm more focused now than I was in those days, although she'd not be pleased with me about all those rest breaks spent sitting on a freshly cut stump out in the woods.

There is a whole order of insects, the mayflies, which we, whose life spans we've made the standard, have named the Ephemerata because we regard them as so short-lived: adults last a day, two at the most. If stones did the naming, we'd be Ephemerata, too. In biological time the individual life, mayfly or human, is not particularly important. Its importance is only to others of our kind.

It doesn't make me sad to contemplate my own time's ending. After all, I've had a pretty good run of it, have done a lot of interesting stuff. I have no regrets, which, as I see people die who have them, I realize is significant. What does make me sad is the accumulating loss over the years of the lives of others dear to me. It isn't an incapacitating sorrow, merely a mild, soft sadness, like some midnight blue velvet background, the setting for new attachments. It gives a luster to the ones I still have, makes me more careful with them. It is backdrop, too, brightening the contrast with new lives coming into being, new shoots of trees coming up, puppies, bee larvae, the family babies. It is cheering to keep the process going.

Toward the end of the summer after the house was finished, Arne came up for a weekend, and we had a party, invited all our neighbors along the road and the people we'd come to know in town and nearby.

It wasn't the same sort of party as my Missouri one. That was an ending; this was a beginning. But I stopped once again and took it all in, took a mental snapshot. There were people crowded in the tower looking out across the ocean; there were people looking at construction details, asking questions of the architect and contractor, both of whom had come; there were people throughout the house in every room, drinks in hand, talking happily. These were people I'd come to know and admire in the years I've been easing myself into Maine. Some of them may become as dear to me as my Missouri friends. A group of children had discovered the loft space created when we'd opened up a part of the former attic, and they were bouncing on the mattresses I'd stored there, giggling.

I'll not live long enough to see those slow-growing Ozark red oaks as stately trees. But those children and our grandchildren, honorary and biological, may. And, after all, that is no small thing. It's as good a reason for planting acorns as any I know.

Index

Page numbers in italics refer to illustrations.